Cover image by David Provolo

AF892566

www.godofhoneybees.com

First edition

ISBN: 978-0-578-57997-9

This book was professionally typeset on Reedsy.

Find out more at reedsy.com

Contents

To my son, Oliver:

I thought this book was for me, but it was really for you. May it serve as a reminder that it is okay to not know, to re-learn, to tear apart and rebuild. If I could choose to instill one thing in you, it is that you would be a critical thinker. Challenge arguments, pull apart concepts and decide for yourself what you think. But all the while, remember this: have fun. All the world is a play and you, my beautiful boy, have a part.

Acknowledgment

I want to say thank you to everyone that I have bugged with questions while forming what is now this book. I would also like to specifically thank Patience Uhlman who runs 29 Pilgrims. Your work on editing this book went beyond simply making sure things made sense. The insight and ideas that you brought to this project helped shape it in a way that conveyed the ideas more clearly than I could have hoped for. Thank you for taking the risk on a religious, nonfiction book.

I also want to say thank you to my wife for allowing me to type in bed, bug her with questions about the cover, and bending her ear with talk about this book.

What is this book?

This book should not be thought of as a prescription because it's not meant to cure anything. Think of this book as a door in the room you're standing in, and that door opens to another room; this might help you understand the motivation behind these words. This book isn't going to force you into the next room. Instead, it will open that door and show you that the building you're in has a lot more space than you may have realized.

I read recently that poet and author T.S. Elliot once responded with words of wisdom to a young girl asking for advice for early writers. He told her to write for herself first (Popova). I feel this is incredibly simple yet genuine advice. What better way to stay true to your most cherished ideas than to write the story you would most want to read? When you let your thoughts wander and begin to take on the burden of the opinions of every potential reader, you may find yourself watering down your purpose and diminishing the glow of raw creation. With this precept in mind, I will stay true to my purpose with this book. Why write this book anyway? I wrote it because I couldn't help the urge. I wrote it because it didn't used to exist, and now it does. I wrote it because we as living beings are on the precise edge of an ever-growing understanding of our world, and it is a beautiful and pure thing to experience. I wrote it simply because it makes me so happy to do so.

Among all the different shades of skin, personalities, and cultures along the human spectrum, we can find many common denominators. One of those factors is our insatiable urge to found and maintain a solid set of beliefs and a clear understanding with which to move through the world. Wouldn't it be terrifying to navigate the world without the words to understand your experience? This fundamental human urge we possess to name and categorize our world leads us to seek out any system that tells us, "this is the way things are." It is in this way that we find ourselves white-knuckling literal, black-and-white views of the world. It's no secret that this drive to understand is prevalent throughout all religious thought regardless of faith or doctrine. In the highly informative book, *A History of God* by Karen Armstrong, I learned about the ways that humans have sought to express the inexpressible (Armstrong). Based on the data we have and our — admittedly limited— understanding of our ancient ancestors, it seems we as humans have always created gods to illustrate that which cannot be expressed in everyday categories. Armstrong explains the way our current state of religious understanding evolved from a long line of iterations. As the group-consciousness matured over time, humans fashioned new ways to express and connect with that ever-present stream of true existence. It took on many forms, from the pagan concept of the great, formless, ocean of creative potential, to sky gods like Baal, Marduk, and

Ashtar, all the way up to modern expressions of the divine. My purpose in pointing this out at the start is to illustrate an operating intention for this book. In the same way that expressions of the nature of the divine have evolved in macro-scale shifts, it also evolves on the micro, individual, human scale. Perhaps the shifts even originate here and move outward, like the ripples from a pebble dropped in a pond. This book is simply an explanation of my expression of the divine. It's an attempt to walk with you through the different lenses I've looked through, these lenses that have come together to form how I understand the world. You reading this book gives me great joy, as a child excited to show his toys to visitors. Thank you for joining me.

The specific faith that I want to talk about is Christianity. More specifically, I would like to address modern, western Christianity. This is because it is the faith that molded my world view during the most formative years of my life. I don't wish to attack the Christian faith using these pages, but rather I aim to point out that in our attempts to categorize, explain, and define our world, we have overlooked some crucial aspects as well as the truth that is all around us.

If we assume that the Christian Bible is to be taken literally and that God spoke to the people mentioned in it, then it paints an interesting picture. Say you have a group of people. This group has been oppressed with the boot of tyranny pressed on their necks and they are searching for a new identity, a new story to tell about themselves. You must tell them a story that everyone in that group can understand. If the people to whom God spoke needed a theology degree or seminary school to understand what was being said, then it wouldn't be much of a story worth telling. No verse says God asked, "Who among you will attend Bible college to make sure you can argue doctrine aggressively?" No, the story was much simpler and thus, more powerful.

Think of it this way: you are trying to tell someone about a poem, and you want to convey the way the poem makes you feel because you sense its beauty deep in your bones. If you start telling this person about the poet's preferred use of semicolons over commas, they'll start to lose interest in what you're saying. The emotion and the beauty of the poem would be diminished by getting caught up in the nitty-gritty grammar and mechanics of it. The beauty is simpler. Similarly, the story told to the people in the Bible is a simple one for that same reason: accessibility. What I want to explore are the foundations of this accessible narrative. Just because a group can adopt a new story about their culture, tribe, homeland, does not mean this is a positive outcome. In much the same way that parent material under the soil influences what kind of soil will be produced, I believe that the origins of this narrative act as the parent material for certain fundamental concepts and do indeed form the collective faith. I want to look at the concepts that form the foundations of the faith, the ones that we in modern practice still adhere to and examine why they may have come about and how it relates to our lived experience in this reality.

What I hope to do is carry on in the tradition of making information accessible to a larger audience. Thankfully modern technology means that a greater number of people are not only literate but also have greater access to an expanse of available resources, so we've already got a leg up. I aim to condense studies and topics into simple concepts that we can examine. By making this information accessible, it will also become shareable, relatable, and engaging.

What is this book then? Well, it's a bit hard to pin down. We are going to start by examining some of the most basic assumptions that I feel come with learning about the Christian faith. Then we will go one layer deeper and explore why these assumptions might not be the whole picture. Then we will explore where this leaves us and what it means for our role in the world. Eventually, I hope to show how this process of examining and questioning the foundation of some commonly-held beliefs leads to a whole new way to see the world. What are my credentials on which I base my argument? I've done a good amount of reading, listening, and studying, and it's led me to where I am now. Aside from this, I have no credentials. Except that I am human. I have lived and suffered and loved and lost. In this way, I hold the same credentials that we all hold and the only ones necessary to discuss the idea of a higher power. I try to explain it from the most open-hearted place that I can and hold it lightly.

So what makes this book different from others like it? I am not arguing the nature of the Christian God from the perspective of religion, but rather I want to explore the nature of God through a lens of philosophy and layman's logic. The Bible seems to be only concerned with the human story. The argument could be made that it is also only concerned with the male story. Other religions, such as Hinduism and Native American lore, have gods that focus on nature, weather, and abstract forces like love and war. I think that the Bible's failings are in its tunnel-vision concern with only humans, and I think that the Bible doesn't go deep enough. Because of this, the Bible, Christianity, and any other religion that concerns itself solely with the human story misses the point. I think that our human story is inextricably intertwined with everything around us. We are a product of the earth and the relationship between two humans and between our planet and the sun. We share this story with every living thing that is around us. The god or gods of worldly religious texts are the results of humans interpreting the divine as it relates to humans. We don't currently understand a god that extends to other forms of life because we cannot extend our consciousness to that life. That is to say, we cannot know what it is to be a different kind of life form. Thus, we do not write our religions to include other forms of life. This book is arguing the nature of God — and I use that name loosely — from the perspective that the concept would extend above, through, and below humans, would extend to every form of life. I'm not interested in drawing lines between one religious interpretation to the next. Circular discussions regarding the afterlife do not help us understand our position in the universe. If anything is clear about humans and our relationship to the environment and our world, it's that we fundamentally do not understand it, and millennia of religious interpretation has not helped us change this. This book draws lines to connect us, all other forms of life, and the God, energy, ground of being, or awareness that lies underneath it all.

I want to clarify that this book is intended only to give you something to consider. This is not an attempt to draw lines in the sand or create another us-against-them argument. This book and the ideas it presents aren't revelatory. However, the ideas I will attempt to convey do seem to be more clearly communicated in this imagery and story that has consumed my mind. This isn't a revelation to new ideas; it's simply reminding us of what we already knew but lost sight of in our human attempts to find security.

Keep in mind we aren't throwing the baby out with the bathwater. We aren't using one conclusion about one particular question as a new pulpit from which we toss the whole idea of religion either. It's so much bigger than that. In my experience, opening one's heart in this way is a bit like getting ready to jump into cold water.

Picture yourself standing on a wooden dock, hearing the lake water lapping under your bare feet. You know that if you jump, the cold water will be unsettling at first, shocking even, but it'll be invigorating once you're in. I hope that by examining, challenging, and learning more about the world around us, people — myself included — can start to hold things lightly. I hope we can take our views about how the world works and loosen our grip on them just enough so that we might really hear our neighbors when they speak.

When I began to write this book, I had one worry right from the beginning: what if the premise of my argument — the religious beliefs or assumptions of my upbringing — were not at all common? It seemed entirely possible that I could take the time to fully explain one point of view only to discover that nobody held the same assertions in the first place.

With that in mind, I decided to consult people I considered to be an authority on a Christian world view. These are people with whom I may not agree but I highly respect and trust. I asked them a series of questions that I felt could derive a decent picture of the world view within which they operate. Of course, my questions were not remotely comprehensive or exhaustive, but I do feel like their answers show some of the leading beliefs that get pulled into common dialogue regarding the subject of Christianity. To any reader that is unfamiliar with the Christian world view, you can take comfort in knowing that the one I critique here is not only the view I was raised with, but also shared by many. Whole congregations gather under the views presented here. For any reader who may be of that congregation, I ask only that you allow for the consideration of the ideas presented here.

You may also notice much of the content in this book revolves around the creation story in Genesis. I have made that decision purposefully because, in my experience, much of the way a Christian worldview works is based on the story in Genesis. Fundamental assumptions of the way the world works, how we should think about God, and the meaning of life are commonly derived from the framework we find in that story.

That being said, I do not claim to argue about the nature of God or gods put forth by other religious texts. I'm not familiar enough with nor a member of a society that unanimously teaches these ways of life in conjunction with Christian teachings. Although, it is arguable that some of the attributes associated with the Christian God mirror other religious deities, so the symmetry may be well-suited. If the concept presented here is at all helpful, then I am overjoyed. If you find it lacking, it very well may be.

Adam and the Donuts

Why do bad things happen? This is the question I would like to start with. I've chosen this question as the starting point because it seems that so much of our mental furniture is arranged around how we answer this question. It affects not only how we view ourselves or how we choose to behave, but it also affects how we move through the world and relate to it. That is to say, the "how" we reconcile the good and the bad directs our lives. I know this question is as old as time and we may never know the definitive answer, and I certainly don't claim to have a definitive one. I do, however, feel I have one that has given me peace.

As I consulted others regarding this question and the ones that follow, I received a decent amount of helpful information. One foundational aspect of the western Christian world view is the belief that God created a world without sin. This belief is illustrated in the book of Genesis in the story of Adam and Eve and the garden of Eden. The story shows a world that is a paradise for all creation. Adam and Eve have everything they would ever need in it, free of sickness or despair.

You can picture the scene described in the book of Genesis. This is the original paradise.

There was, however, a kink in the chain. This kink is the interesting relationship between creator and creation. God created Adam and Eve and Lilith, and in doing so, created their limitations. For those that may not have heard of Lilith, as I hadn't when I started writing this book, her existence serves as a blatant and the earliest inconsistency in the Bible. There are numerous examples where the writer (or writers) betrays the very human aspect of what is written in the Bible. Lilith is one such example, so let's begin here — at the beginning.

In Genesis 1, we find a creation story that moves through the myth in a way that we are familiar with: dividing notions like light and dark, creating creatures like birds and fish. Then we come to humans where the plurality of that word is the key concept. God says mankind should be made in "our" image, so he creates man and woman and then instructs them to repopulate the Earth. If this strikes you as odd, join the club. This passage is one that I took for granted growing up in a private, Christian school and the associated church. When I re-examined the chapter, I was shocked that I never noticed this before. Doesn't this run counter to the idea that Adam was alone for a while and Eve came because Adam was lonely? This is what I had always been taught in school, Sunday school, church, etc. Let's move to Genesis 2.

In Genesis 2, we find a very similar creation story including dividing and separating and creating things. However, when we get to the human part, it's a very different story. The singularity of the term *human* here is the key concept. God creates Adam, and that's it. It is suggested that Adam is alone for a long, long time. He wanders the earth looking for a partner and giving names to what he finds. It's only once God notices that Adam can't find a suitable partner in all the animals that have been created, that God gets the idea to create another human. We have come across two aspects of this story that would run counter to God's omniscient nature. The first contradiction being the notion that God did not know ahead of time that His creation would grow lonely, which would undermine His omnipotent nature, but we will examine that more closely later. The more blatant error is that these two stories exist back-to-back. Their chapters are one after the other.

This posed a particular problem for early followers. To marry these two stories and smudge out the contradiction, early writers developed or borrowed, the character named Lilith. I say borrowed because the myth of Lilith has its roots much earlier in human history, around 3000 BCE (Hammer) than the writing of Genesis, estimated to be written around 1445 BCE (Stewart). The archetype that shows up in so many early myths of a female entity or she-demon representing darkness, terror, sensuality, and unbridled freedom, is also found in the Talmud. If you simply read the first and second chapters of Genesis, you will see the contradiction, but you won't find the story of Lilith. You would have to find the Book of Ben Sira to find the tale. According to the Book of Ben Sira, Lilith was Adam's first wife, but after she refused to be submissive to him, she went out on her own, became a demon, and moved to the sea. All this narrative shift just to reconcile the two opposing stories. Now, back to Adam and his *second* wife, Eve.

Adam and Eve know nothing but love, happiness, and health. Their world is pure, and they are not yet tainted with sin, therefore they do not have a choice *not* to love God. A world in which the creation doesn't have the option to *not* love the creator is not a world of true love. Love must be freely given and accepted. It is something that every human experiences, be it toward our parents, our children, our spouses, or our pets. Love is a dance in which we bend and twist and twirl together. As soon as one begins to force the movements of the other, the dance falls apart. Adam and Eve in this paradise would be no more than fleshy robots simply carrying out protocol programmed by the Creator. We will get into the gift of free will later, but for now, we just need to know that to resolve this lack of true love, God gave the gift of free will to Adam and Eve for them to be able to *choose* to love Him. This notion of God granting free will for us to have the opportunity to choose to love him happens to be, in my opinion, one of the fundamental concepts of Christianity and forms a basis for the reason followers choose this spiritual path.

Hold this lightly. Let's just imagine. Given the picture of Eden and the two humans, we are left with two questions. These questions form the basis for some of the concepts presented in this book. Question one: in a world where there is no sin — by which we mean anything that goes against the perfect blueprint that God created — how can Adam and Eve have any alternative other than to love the Creator? This is the question we will look at in the next chapter. Question two: where did that alternative — which we call evil — come from? Where did the antithesis to an all-knowing, all-powerful creator come from?

God created the world in total perfection. His last creation was man and then later, woman. Adam and Eve existed in a perfect world and in alignment with the Good — with a capital G — that God intended. The only rule was that Adam and Eve were forbidden to eat from the Tree of Knowledge of Good and Evil. Keep that name in mind for later: Knowledge of Good and Evil. Up until the point when the serpent (the entity we will address shortly) enters the scene, the two humans are never mentioned to have the temptation to eat from the tree. This is an important point. To illustrate how interesting this arrangement is, let's look at an example.

Imagine a windowless room with only one door that stays closed. Inside the room, there is only a bed, a table with food, and a child. The child has only ever existed in this room and has no concept of any existence that may be on the other side of the door. Let's give the child a name to make things easier. How about Adam?

Adam, in this room, never gets too cold or too hot. He is never too hungry or too thirsty. He is never uncomfortable, never gets sick. On the table is food but only the essentials to stay healthy and grow. The thing about this arrangement is, however, that right on the other side of that door is another table full of donuts. Sweet, warm, donuts.

Now a question. Does Adam, in this scenario, have free will over his life? Well, so far as the room is concerned, yes, he does. He can choose to not eat the life-giving, healthy food on the table. He could pick and choose which foods on the table he likes best and he can avoid others. He could adhere to a specific kind of self-inflicted diet or gorge on everything in sight. The problem is that he doesn't have the full picture. These donuts exist on the other side of that door, but Adam has no window through which to see these donuts, no opportunity to even see that he has a choice of cuisine. How can he truly decide to only adhere to the life-giving food when he doesn't know that there are donuts right outside? Furthermore, he doesn't even know what "outside" is. How can Adam have free will if there is no other choice? Isn't free will our ability to make our own choices?

Now take adult Adam and place him in the Garden of Eden having just come into conscious awareness. All Adam knows is what's around him and how he feels. If God created everything in perfection, then Adam does not know anything aside from this perfect reality. We have to question whether or not Adam possesses free will if no antithesis to the Creator exists anywhere. One may argue that Adam could still reject God, but how? Reject him for what? The very concept of anything going against God hasn't entered the picture yet. What other paths could Adam choose? There is no concept of outside-the-room.

Enter Lucifer. Lucifer — also referred to as Satan, the Devil, Beelzebub, and many other names — is the one entity to which all evil and bad things are attributed. We assume he entered the garden of Eden as a serpent, but let's go back to his creation and start there. Before the Earth and all of God's creations existed, it's believed that there was only God and the Angels. The definitive answer on what came before God created everything is still hotly debated, but if we are going to look at this through a literal interpretation, many come down on the side of Lucifer being banished before the creation of Adam and Eve. That is to say, the story we are familiar with regarding Lucifer being cast down from Heaven is largely assumed to have taken place before the creation of the original humans.

If we adhere to the idea of a God that created everything, then we can assume that God created the Angels as well. If you are one of the Angels hanging out with God and there is no other creation to speak of, it would be pretty obvious where you came from. If you come into existence and God Himself looks you in the eye and informs you that He created you, you wouldn't have much to argue. We don't have much to go off in the Bible as to what existed before God created everything, nor do we know what this proverbial middle-ground in between "before" and "after" creation was like for these Angels. However, the state in which God and the Angels existed does kind of sound like the Garden of Eden, doesn't it? There is an all-knowing, all-powerful creator that communes directly with you within the reality He created which itself is pure perfection.

Again, where is the concept of evil going to come from in an environment like this? We have two choices: either God created evil or Lucifer possessed free will. To decide between these two choices, we must define the three attributes of God that are widely accepted: He is omniscient, omnipotent, and omnipresent. This means that God created all, knows all, and is always present everywhere at the same time. If we accept that God is omniscient, omnipotent, and omnipresent, then we accept that He must know exactly what will come from creating Lucifer. Now the question remains: did God give Lucifer free will? If He did, and Lucifer chose to rebel against God, then it still follows that God created something that He knew would rebel. Regardless of which side you take on this, one thing is clear: God would have had to create the concept of His antithesis.

Think carefully about what this would mean. The all-knowing God creates an entity that He either *knows* will become His opposite, or He is going to take the chance. Keep this in mind the next time your stomach turns at the news of some tragedy. Either way, God is knowingly creating His nemesis, and to do that, He has to create the framework for that nemesis to operate within. Otherwise, the entire concept of good and evil exist outside the context of God and Lucifer, and they are playing by someone else's rules, thereby making God not all-powerful or the source of all good. Isaiah 45:7 says, "I form the light, and create darkness: I make peace, and create evil: I the LORD do all these things" (King James Version, Isa. 45:7).

I don't want to just throw verses around to try proving a point. The Bible, as with most any text, can be taken out of context to prove just about anything. So with the utmost respect to the Bible, let's discuss this verse in context, the way it would have been understood when it came about. The book of Isaiah is interesting because, among other things, there is great debate as to whether or not it was written by only the single prophet due to the immense time frame it references. For our purposes, though, we will be focusing on the context of this one verse. In the above quote, the writer is speaking to Cyrus the Great who ended Babylonian captivity (King James Version, Isa. 45). Throughout the chapter, the writer describes the God that will help Cyrus in his battle.

Some have argued that the original Hebrew word *ra* — translated in the above quote as "evil" — does not necessarily mean a moral evil or an evil brought about by humans. In this view, Isaiah 45:7 would mean that God brings calamity in the form of judgment upon sinners (Slick). This view also argues that other verses could show that God cannot approve of moral evil. However, this doesn't put our question to rest, does it? This view also happens to be wrong. The word "ra" can mean everything from natural disasters and calamity to a moral evil like murder. This means that an attempt to justify this verse by saying God has nothing to do with moral evil is whitewashing God's character. Remember the name of the tree from the garden? The Hebrew name is "tov wa-ra" translating to "good and evil." There are two important points regarding this name. First, it contains the word "ra" that we have shown includes moral evil. Second, since it is the tree of good and evil, or left and right, or any other dichotomy we may choose, it could be argued that it contains everything in its name. It could be called the tree of everything. Aside from the errors and inconsistencies in this creation myth, what a perfect name for this tree. If we think of Adam in the room, suppose the whole rest of the world is on the other side of that door. Everything is on the other side of that door. Once Adam is exposed to the outside world, the donuts serve as motivation to engage the rest of existence, to open up to the other half of experience. In Eden, once Adam takes a bite of the fruit of this "everything" tree, it opens him up to the rest of conscious experience. He now has the other half of life, instead of the limited environment and choices bestowed upon him by God. Perhaps now he has free will since he is seeing the full spectrum of options he can choose from. It made him "like a God" because he held the full range of experience in his mind. Good name choice.

The question remains: where does this "moral evil" come from? If we think about the question using our earlier Adam experiment with the donuts, it seems that the Creator would have to create His antithesis for it to exist.

Okay, so given that God would have had to create His opposite and the framework in which the opposite would operate, the next question is why? Why create such an adversary and then allow him to enter the perfect garden paradise you just created? Let's return to the garden.

Adam and Eve are existing in a literal paradise. What do you picture when you think of paradise? Take note of your paradigm of the word "paradise" because it's likely not big enough to hold this view of Eden. It's not just your favorite vacation spot. In God's paradise, Adam and Eve never get sick or hurt or too cold or too hot. Exactly like the "Adam and the donuts" analogy from before.

If Adam and Eve were to remain existing in Eden in this way, free of anything that was not purely perfect from God, they would be like Adam in the room and would not truly have free will. If we assume that a loving God must give His creations free will so that they may *choose* His path, then we can see why God had to allow something impure into Eden. Adam and Eve need to be, dare I say, tempted by an alternative. How else could Adam and Eve choose?

However, that means that God had to allow Lucifer to enter Eden. This means that God didn't protect the most perfect human beings to ever walk the Earth from "evil." Furthermore, this means that God didn't protect the most perfect human beings from an evil that He essentially created. Weird.

One more point about Lucifer. Nowhere in the Bible does it say that the serpent was Lucifer. The concept of the Devil as a separate, independent entity did not exist in ancient Israel's understanding. This idea of Satan as the snake came about much later and was born in response to the tumultuous existence within the Jewish community. Shawna Dolansky, a research professor in religion at Carleton University, explains this seemingly sudden transformation in great detail in her article "How the Serpent Became Satan." In the article, Dolansky explains how the satan — not Satan as a proper noun — was used as a tool of God and in heavenly council. I agree with Dolansky and her assertions. But whether or not you believe that Satan was a living, talking being or simply a tool used by God, the mere existence of the controversy as to whether or not Lucifer was ever even in the garden shows that our modern understanding of Satan is misguided.

The inconsistency in the narrative of the foundational story is enough to challenge how literally we can or cannot take the Bible. The fact that the Bible isn't clear on this point (and that scholars studying the origins of the story don't agree) means that we should approach this tale with a more open hand.

Let's take the "Adam in the room" example one step further. Imagine the door opens and Adam smells the warm donuts on the other side. Not only is he presented with a major change to his reality, but his senses would be, if nothing else, attuned to and perhaps overwhelmed by the new smell. If he walks through the door, finds the donuts, and decides to take a bite of one, should we hold him responsible for doing something wrong? Are the donuts evil? If so, based on what framework?

In one sense, we would have to say no, we can't say he did something wrong. Why? Because Adam didn't know what donuts were or why you should pick the healthy food on the table instead of sweet, fatty, fried dough. He didn't even have a paradigm of what the word "donut" is connected to. His actions and decisions are predicated on the environment and experience his psyche holds. This concept is explored more deeply in the book *Free Will,* a piece written by neuroscientist and best-selling author Sam Harris. Here, we can see another point where the writing of the Bible betrays the very human and time-specific mentality of the text. We can see clearly that Adam in the room did nothing wrong by eating the donut. Eve, existing in a realm where there is no concept of not-God, could not be accused of doing anything *wrong* by eating the fruit. She doesn't have that mental category yet. Before eating from the tree, Eve heard that she should not eat from it or else she would die. This is a threat that makes no sense without the concept of death. What does this leave us with? There must be a separate motivation for writing not one, but two female characters as the temptress, the representation of morally weak, the lesser.

What's interesting in this scenario is that once Adam is in front of those donuts, nothing has changed for his experience. Once he steps out of that door, the room he *was* physically inside of becomes the room he *is* mentally inside of. He is still Adam in the room. It is with this framework that he will filter the world. We could argue that the donuts are serving as the temptation to draw him out of the room. The growth of his psyche is not only a new taste experience but that his world or environment is larger than he had thought.

Can you see why evil must exist? Once Adam has learned that more options exist, that there is more to his environment than he previously thought, he does have the option to go back to his room and shut himself in. This demonstrates, if only in a very simple sense, free will. It also demonstrates why evil must exist, why growth and beauty and death and *ra* are intertwined. If Adam's door was locked and never opened, how truly free could we say he is?

What's more, can you see all the questions and scenarios that don't make sense when you force an origin story like Adam and Eve to be literal? Did God always intend for Adam and Even to eventually die? Why would He not guard Eden from the serpent? Why put a tree that can ruin everything right there in the garden?

After all, the first verse in the Bible doesn't include a line that says, "in the beginning, God created pencil and paper and wrote down what happened."

The reason there are so many logical inconsistencies in the book of Genesis is not because the writer was failing to tell the story with accuracy. These flaws come to light when we put a certain lens on the story. A lens that was not meant for this tale. The creation story lends itself to our critique when we try to draw the meaning of life from this small story. Many scholars agree that the book of Genesis was never meant to be a literal creation story. Bruce Waltke, an evangelical scholar who has a master's degree in theology and a doctorate in theology, as well as Jon D. Levenson, a professor of Jewish studies at Harvard Divinity School, are among the many critics that point out that Genesis wasn't meant to be a historical account of the creation of the universe. They argue that the creation story in Genesis is mainly a critique of Mesopotamian theology which asserted that creation was born from violence among many gods and that this story of creation lies at the heart of reality. Genesis, written by an enslaved people far from home, asserts that the world is fundamentally good and born from perfection and love. Also that it came from one source, not many.

The fact that the story of Adam and Eve was written as a critique of Mesopotamian theology is no reason to disregard the story entirely. It's only a reason to do away with a lens that may not be appropriate for examining the story. Genesis does not seem to teach that God gave humans an inch and they took a mile, and as a result we now all have to pay our dues, tisk tisk. What can be seen is that even in the paradise that God created, Adam and Eve did not truly have free will. It is only once the serpent came in and expanded their paradigm to include the other side of the coin did they wake up. There are many things that the creation story in Genesis can teach, but I would argue that one of the most important lessons is that Good and Evil bump up against and support each other in a dance that has gone on for as long as we know. The question then is not "why should we be compelled to do good if good and evil balance each other out anyway?" because we know why. Christians, Atheists, Buddhists, Muslims, and every single other human mental framework feel the pangs of witnessing evil in the world. We know what kind of world we want. It's just a matter of what kind of world are we going to work toward. We, being awake to the nature of our world, awake to the existence of both sides of the coin, can have our choice, as Genesis points out, but the choice is ours no less. This does not undermine the story of Adam and Eve. It allows it to stretch and grow and shine. In that light, we can begin to see the roots of this tale stretching far into the history of humans.

Age of the Automatons

What a sticky subject free will can be. The concept can become blindingly murky when you begin to consider the near countless views put forth by genius philosophers. I don't intend to argue for the validity of any view in particular or rehash all the complex points presented by most. What I do want to address is how one may think of free will in the everyday sense or layman's terms. Seminary students may be screaming at my book right now (if they even pick it up in the first place), but if this topic is going to be accessible, as it was at its origin, then we won't spend time splitting hairs over dense frameworks however useful they may be. Given our examination of the garden of Eden in a literal context, we can assume that Lucifer entering the garden was God allowing for His creation to choose His path rather than being forced down it. Let's again ask this question: Why do bad things happen?

According to those from whom I sought counsel regarding the Christian view on this, rejecting God leads to the world being out of balance and descending into chaos and evil. In other words, whenever we go against God's will, we begin walking the path of evil, calamity, and sin. I heard it argued many times that this simple premise is the root of evil in the world. Interestingly, the assumption that the devil is the reason for bad things happening in the world is a concept presented in a cognitive test. In a study that looks at the relationship between fundamentalist religious beliefs and the acceptance of fake news, the authors reference a scale used to measure a subject's degree of religious fundamentalism. This scale presents the subject with a series of statements to which they rate how strongly they disagree or agree. One statement presented is this: "The basic cause of evil in this world is Satan, who is still constantly and ferociously fighting against God" (Bronstein et al.). A strong degree of agreement with this statement is correlated with a fundamentalist religious outlook.

Holders of this view might long for a time since passed when society focused more on God and may blame the dissolution on secularism pulling that focus elsewhere. This view can be categorized as a cognitive bias known as Declinism, or the assumption that society is tending toward decay, or Rosy Retrospection, the assumption that the past was a better time. Either way, the idea is that what's wrong with the world today can be blamed on society pulling away from God and His will. At this point, a possible answer to our question, why do bad things happen, would be that they happen whenever we diverge from God's will.

Now let's relate why bad things happen to the concept of free will. Doing this tends to lead many to the conclusion that bad things happen as a result of God-given free will, and if we would just obey His way, then the problem would be solved. Although God can't just force us to obey because then we would be just human machines taking orders. I'd like to ask: would we?

The idea of God forcing us to choose His path for us, I feel, can be viewed in two ways. In one view, God's will is a bit like mind control where God actively influences our conscious thought. This would mean actively interrupting our free will. In the other view, God created an environment where choices that are in opposition to God's plan don't exist. In both of these hypothetical scenarios, the individual would exist on a one-track existence. The single option that would exist in either realm is forced adherence to one plan, one will for humanity. Let's assume the latter: a version of reality where those choices that go against God's plan don't even exist.

Close your eyes and picture the world around you as it is today. In your current place and current time with all the friends, family, and neighbors in your life. Also, picture everything that's wrong with the world around you. All the pain and suffering you see, all the evil that has taken place throughout human history and all the evil that takes place now. Now with this in mind, strip away all the pain and suffering and evil. Picture only the things that you would consider good, beautiful, and perfect. Imagine that the opportunity for someone to choose to harm someone or engage in evil ways isn't even an option. In this version of reality where only good things exist, why would anyone choose anything else? More to the point, *how* could anyone pick anything else? If murder, for example, didn't exist as a reality, then we couldn't exactly choose to murder. What would it feel like to live in that place? Would it feel like a utopic freedom or a dictatorship? I would argue that it would feel a lot like Eden. It would feel a lot like Adam in the room where he is always content. And contentment is subjective because you need an alternative to *not* be content to have a reference point of what it is to *be* content. So in this utopic world where there is no evil, would we truly be happy? Would we know what good is if we didn't have the alternative to compare it to?

This hypothetical picture of reality is the reality that Adam and Eve would have existed in. No, it does not include the tree from which we are not to eat, but there is no mention in the Bible that Adam or Eve had any struggles with temptation regarding the tree until after Lucifer showed up. So just like Adam in the room, we in the utopic paradise would still have free will to choose the options presented to us, just not anything else. We can see then, that the topic of free will is largely dependent on context. In one context we, of course, have free will to make a selection among the options presented to us. In another context, we do not have free will to choose the options that aren't presented to us. Freedom or dictatorship? Free will or control?

Let's take it a step further. In our hypothetical scenario, not only has God created an environment in which we do not have the opportunity to disobey Him, but He is in control of what happens to us. He is the all-powerful conductor of the events that occur in our lives. All, supposedly, working together for the good of His followers.

How should we think about or categorize a being that is in complete control of the events in our lives and is all-powerful and all-knowing? There are countless and endless debates over the concept of an omnipotent deity. The problem with these arguments, it would seem, is the logic and circular reasoning one falls into while trying to define and defend such a deity. I came across one such discussion in which the speaker presents logical arguments around defining omnipotence. The blog's author, going by the name AnticitizenX, demonstrates simply and clearly how the idea that omnipotence means being able to do anything, and that it commits a logical fallacy and contradicts itself (AnticitizenX). One aspect of AnticitizenX's approach that I enjoy is that he doesn't get caught in the weeds of later problems of omnipotence. He starts at the beginning and won't look past mental allowances one may give to reach later arguments. Initially, the logical error rests in language. The example he provides is creating a married bachelor which, of course, can't be done because by definition a bachelor isn't married; it's an oxymoron, a linguistic contradiction. While this may not serve as an argument against the existence of an omnipotent deity outright, it does illustrate a fundamental issue. That issue being that language limits our categories for thought or experience (AnticitizenX). If the deity were able to create a married bachelor, then a married bachelor would have to exist outside the context of language and therefore could not be categorized as a married bachelor. Simply put, we wouldn't have a frame of reference to experience a married bachelor because we would have to drop all words and concepts that we use to understand the world. Once AnticitizenX has explained how the definition of omnipotence meaning "being able to do anything" is faulty, he then presents an alternative definition. This one says omnipotence means being able to do all logically possible things. The adjusted definition gets around the contradiction presented in the first definition, which happens to be the definition argued by the Bible, however, it eliminates the Christian God with a simple thought experiment. Is it logically possible to tell a lie? Yes, of course, it is. This means that an omnipotent God would have to be at least able to tell a lie if He were to be omnipotent. However, God can't tell a lie as told in Hebrews 6:18 "It is impossible for God to lie" (King James Version, Heb. 6:18). Where does that leave us? It leaves us in a tangled mess of semantics while trying to defend the purely good and all-powerful character of God.

I think the reason we end up in these circles and semantic battles is because we are guilty of attempting to whitewash God's character and we are afraid to admit something to ourselves: that good and evil are of the same origin, dance the same dance, and this world will always require both. Once we recognize the inconsistent logic of an all-powerful creator being pure goodness, and the Devil being pure evil, and we see that they are two sides of the same coin, we can begin to reconcile these opposed concepts. I think the natural world frequently illustrates truths about the fundamental nature of reality, and this scenario is no different. What we eat and how we relate to that food can also shine a light on the false dichotomy of good and evil being mutually exclusive.

One of my favorite podcasts, *Radiolab*, had an episode, "From Tree to Shining Tree", that discussed how trees communicate with other plant life in response to stimuli in their environment ("From Tree to Shining Tree"). The episode also showed how plant life exists in an interconnected subterranean communication network. The trees share information about threats to their environment, they share resources, and they respond to their surroundings. While trees don't have a brain in the typical sense, they do display some compelling characteristics of sentient-like organisms. What more must an organism be able to do to be considered "alive"? What does this mean for our diets? We can make sure we never eat an animal product, that our makeup isn't tested on animals, that wildlife habitats remain undisturbed. We make sure we eat only fresh fruits and vegetables and grains. However, what are we supposed to do now that we know the flora all around us is more alive than we previously thought? It is alive and not just in the sense that it is, say, for example, growing in response to sunlight. Are we to make a judgment call on what kind of alive-ness is more worthy of existing unhindered? What are we going to eat if all the food is alive? We can see the raw truth of existence in something common that we take for granted.

Ideas like this, while it is very simplified, bring us face-to-face with the truth that we all know but we spend our lives trying to disprove, rationalize, and pray our way out of — the truth that this world is brutal. Eat the living to stay living, then eventually die and return to the dirt that helped produce you. We can only start to truly reconcile our position in the grand symphony of creation if we accept that good and evil are the push and pull, the plus and minus, the creative and destructive requirements for the kind of world we exist in. Another way to conceptualize this false dichotomy between good and evil is to consider the natural process of reproduction.

Darwin suggested that three factors give way to evolution (Quammen). The first piece was a continuity within species by way of hereditary traits. This essentially means there is a resemblance between offspring and their parents and grandparents. Descendants of a certain species will continue to look, more or less, like that species. The second factor is that while there is a continuity among species, variation does occur within the species. Reproduction is not strictly exact. This means that you are not a genetic clone of your grandmother or grandfather. The final piece is that population growth always outpaces the means of support from the environment (Quammen).

Take for example a spider or an octopus. They will lay numerous eggs with the potential for many offspring. If every one of those offspring were to survive, the world would very quickly be overrun with spiders and octopi. Nature's response to this (using the term "response" at risk of anthropomorphizing a natural process) is not allowing for every offspring to survive. Whether by DNA, predator, or environment, some number of the descendants will not make it. When we consider the kind of world God intended, the original plan before evil entered the world, we may imagine one without death, without tears or pain, etc. But that kind of world simply cannot exist in one where beings reproduce. The population of every reproducing organism, or even of just the soul-bearing ones (which is an argument unto itself), would very quickly outpace the resources and physical space of the surface of the globe.

If God created man and woman and instructed them to populate the Earth with their God-given sex organs, then death and tears and pain were part of the scheme from the beginning. According to the story, Adam and Eve must have been created with sex organs as it doesn't allude to the addition of sex organs after the "fall." Additionally, God commanded them to repopulate the Earth before the tree incident. The order of events would mean that reproduction was the plan all along. Once again, we have an example of putting the wrong lens on the story. Do you see how the story of Adam and Eve breaks down when we assume it to be literal? Forcing a literal reading of a mythic tale will always make the tale fall apart. A positive consequence of forcing this lens, however, is that we can see that this story cannot be taken literally even if it was meant to be considered in this way at the beginning.

For now, just keep in mind the examples presented here: all living things are displaying a level of conscious awareness and the three factors of Darwinian evolution. When looking at the world through this lens, it starts to make one question the distinction between Good and Evil. If we believe God is an entity that is separate from His creation and chooses to get involved when He wants, then it seems like whether or not we get stuck with our current world or the one we imagined in our previous example where the option to choose evil doesn't exist, boils down to God's decision to allow Evil in. Free will void of bad options sounding more like Eden before Lucifer starts to blur the line a bit.

Hold on to this idea as we move forward, just do it with an open hand.

Streets of Gold or Dirt?

Here is what we have established so far: God created a world of total perfection in which the two original humans, Adam and Eve, existed free of suffering. For these two humans to have the opportunity to freely choose God, they must have had free will. Using this God-given free will, Adam and Eve chose to go against God's wishes and thus brought Evil into the world. In this way, suffering became part of our existence whereas it had not been before.

At this point in the story, there is another question: Why is God concerned with whether or not we follow His commands?

God gave free will to the two most perfect humans ever created, and doing so allowed them to reject His love and will which meant that God had to let pain and suffering become part of everyday life from there on out. But why did it matter whether or not Adam or Eve chose to go against His wishes? Many, if not most, western Christians believe that it matters because this has to do with whether or not we will spend our eternity in Heaven with God. This point is very commonplace and is therefore obvious: adhere to God's will and you will be in Heaven; don't and you'll end up in Hell.

I don't wish to wade into the debate over the existence of a literal Hell or the legitimacy of a god that would banish his creation to endless torment, though that is a debate worth unpacking. I only aim to focus on a specific point. The point that God, in this interpretation, cares very much that you adhere to His will so that your soul may spend your eternity in Heaven with Him.

This is an interesting point because based on this reading of the Bible, it is implied that God is predominantly concerned with the afterlife of humans and not any other creation. So this begs the questions, why humans, and what is Heaven?

First, why humans? Why is God so concerned with the choices of Adam and Eve but not anything else? The dominant belief in Christianity is that humans are something unique among creation as we have the "breath of life" giving us a soul, unlike other sentient beings. This soul is the part of us that will be subject to either Heaven or Hell in the afterlife.

It isn't necessary to delve too deeply into the nature of Heaven and its characteristics mainly because all the imagery in the Bible is largely metaphoric, so your paradigm will do. According to Christian doctrine, we as humans have souls, unlike any other creation, that will spend forever with God in this idea of Heaven if we obey Him. We've already established that God created Adam and Eve as the two original humans inside Eden which itself was created in total "good" perfection. Here's what doesn't seem to add up: that very perfection

Eden was created in total perfection, and Adam and Eve were created as perfect human beings intended to commune in paradise with God as they did before the snake entered the garden. Why does God care whether or not these two choose or reject him? See, if Eden and its occupants were perfect and God walked among them in this creation, why then would Heaven be necessary? This leads us to the second question. What is Heaven? If Heaven is a place of eternal bliss, goodness, and perfection, then does that mean that Eden was not a place of utter perfection? Why would death as a prelude to the trip to Heaven be necessary?

The concept of Heaven and Hell (and the contingency plan for the humans to fail) would have to exist *before* the creation of Adam and Eve to justify God giving them free will to choose to accept or reject His love. Was God making a backup plan for the event of human failure? Of course, He was. The afterlife, free will, and having a soul are all tied together. If there was no afterlife, there would be no reason for Adam or Eve to have a soul. If they didn't have souls, then there would be no need for the tree in the garden or the choice to eat the apple. The story that is told about the creation of Eden and Adam suggests that God had planned some form of an afterlife before creating Adam since he was given the breath of life upon his creation.

Odd.

Even if God was just creating a backup plan in case Adam or Eve backslid, why would there be an idea of the humans spending eternity in Heaven or Hell? God created Eden in pure perfection — arguably a heaven — so if anything, shouldn't that necessitate only the existence of Hell as a punishment? Was this fifty-fifty afterlife the plan all along? When looking closely at the book of Genesis, one can see that there is no mention of Adam or Eve existing in Heaven or Hell. Let's take a moment and look closely at the story of Genesis and the account of creation. In Genesis I, verse 29, we find a line from God where he explains what is food and what humans have dominion over.

> Then God said, "I give you every seed-bearing plant on the face of the whole earth and every tree that has fruit with seed in it. They will be yours for food. And to all the beasts of the earth and all the birds in the sky and all the creatures that move along the ground—everything that has the breath of life in it—I give every green plant for food." And it was so. (NIV, Gen. 1:29-30)

Then chapter one ends with the author explaining that God was pleased with all that had been created. As we already know, there are two creation stories in Genesis. How do they compare? Once Adam was on the scene and God was explaining to him what to eat (much like chapter one) there was a caveat added that didn't appear in the first story.

> The Lord God took the man and put him in the Garden of Eden to work it and take care of it. And the Lord God commanded the man, "You are free to eat from any tree in the garden; but you must not eat from the tree of the knowledge of good and evil, for when you eat from it you will certainly die. (NIV, Gen. 2:15-17)

Two points here: we can see the stories don't match up and that Heaven and Hell were not part of the picture. One point that we have come to, and will continue to do so, is that the mere existence of two contradicting stories in the text is reason enough for the reader to take it with a grain of salt. The more pressing point here is that Heaven and Hell were not part of the picture. Adam and Eve were not told they would be sent to Hell if they disobeyed God. They were only told that they would die. This implies that if Adam and Eve had made the "right" choice, the earth would have been tended by two immortal beings forever. We will get to the problem of this framework for Eden later. Further into the story of Adam and Eve, after the deed is done, they are told how much more difficult their existence will be. If Hell was not a threat and Heaven was not a reward, then what in the world was the "breath of life"? The dominant understanding in the religious circles I grew up in was that the breath of life was a way of saying "soul."

Could the breath of life that God gives Adam but not Eve in chapter two, verse 7, have been God giving Adam a soul? The verse in chapter 1, where the term "breath of life" can be found, also happens to put nearly everything God created in the category of having that breath of life. According to an article by Rabbi Jack Abramowitz, the word used in Genesis is "Nefesh." Abramowitz states that this refers to Adam's soul which differs from Adam's physical form (Abramowitz). The word nefesh is used both in Genesis chapter 1, verse 30 as well as chapter 2, verse 7. The first context is concerning the nature of plants and animals, the second is concerning humans. Based on our understanding that a soul seems inherently linked to an afterlife, why does the rest of creation also have a soul?

With no mention of the framework of Heaven or Hell, what purpose does the soul serve here? The original plan, according to chapter one, was for Adam and Eve to go on tending to heaven-on-Earth for all eternity. Or according to chapter two, it was for Adam to do this until Adam got lonely. Either way, the original framework did not include plans for Adam and Eve to die, so who needs an eternal soul? Again, Abramowitz claims that the soul is something that exists independently of the body and it is the way a person can comprehend knowledge beyond the physical. I would ask why they need a special part to understand God when he walks among them in the garden. Abramowitz also claims that *nefesh* will not die when the body dies but instead it lives forever. Yet this explanation doesn't solve the root issue here. Where will the *nefesh* live forever? In Heaven with God? If so, we are back to square one. Further, why would God create a being that has two seemingly independent aspects of their existence, one of which does not perish? The problem is that God seemed to only threaten death if they disobeyed, and death was never part of the deal before the temptation at the tree.

We have already discussed the unique relationship between Lucifer and the garden, but there's more. It's believed by many scholars that Lucifer, whom many believe was the serpent that tempted Eve, was already on Earth before Adam and Eve were created. When Lucifer was banished from Heaven, God cast him "from the mount of God" (King James Version, Ezek. 28:16) and then threw him to the Earth (King James Version, Ezek. 28:17).

The first question is: if the Earth was created in total perfection, what kind of banishment is that? It would seem that the only thing that Lucifer is going without is the adoration of all angels or the immediate opportunity and proximity to try to overthrow God. Aside from that, Lucifer was banished to a literal paradise.

Hold that thought. Lucifer gets banished to Earth before Adam and Eve being created. So that means that God brought the original, perfect human beings into existence in a garden where He just sent His arch-nemesis. That's not all. God then allows the nemesis to enter this garden to tempt those two humans.

That means that Lucifer would be, in this scenario, a tool of God, an instrument, something that He uses purposely. If not, then God surely would have guarded the garden against His nemesis to protect the most perfect humans to ever exist. Dolansky augments this concept when she explains how the story of Job clearly illustrates the relationship between God and the satan, not Satan as a name. If you're like me, you may have pictured a black-cloaked entity slithering up to God's feet and trying to coerce the Creator into a challenge or a bet. While the wagering aspect is accurate, in the actual story of Job, it's laid out quite clearly that the satan will only act on God's instructions and is "not an independent source of evil" (Dolanksy).

All this to say that God is concerned with whether or not we choose His path so that we may commune with Him for eternity in Heaven (as far as a literal reading of the scripture goes). A literal interpretation seems to have some deep questions surrounding it, but to progress to the next point, just keep in mind that this is the motivation behind God desiring us to choose His love.

Unfortunately, for the literal interpretation, this fundamental concept of God being concerned with whether or not we choose Him regarding our afterlife is found to be faulty due to some other content in the Bible itself. In the chapter called *Unforgiving* of *God, The Most Unpleasant Character in All of Fiction* (Barker), all of the verses dealing with the wrath God against people all occurs while they are still alive. People experience His wrath as they are living right now, not in the afterlife. Many readers familiar with the Bible will know there is no shortage of carnage, war, violence, and other woes in the text. What may be surprising, though, is that upon close inspection, the reader can see that the God character is often the one bringing about these calamities (recall ra being used to describe both calamities and moral evil) and they are often against people to get them to do something while they are still alive. God doesn't sit back and let the humans act on their own free will and wait to judge them on the other side. John Drummond, a researcher at Indiana University East, focuses on — among other things — comparative mythology and its influence on the Bible, and he asserts this concept in his article "Who Is Satan?":

> In the Hebrew Bible, [God]'s greatest enemies are not fallen angels commanding armies of

> demons, nor even the gods of other nations, but, rather, human beings. It isn't the devil that spreads evil across the face of creation—it is mankind. (Drummond)

This notion of being spared of God's wrath in the afterlife because you turned to Him while on earth is a moot point. God won't even spare wrath on you *before* you die! I have read and heard arguments to this point that suggest that God's wrath was quenched with the sacrifice of Jesus. I take two issues with this argument. First, the foundational ideas about God and the afterlife are built on sand. The contradiction and logical errors that are made are enough to question even the essence of the concept, never mind the latter ideas. Second, this response that Jesus quelled a raging God means that God Himself is a terrible being and what humans really need is Jesus, not God. Humanity needs Jesus to battle God for the right to be our leader. However, since Jesus is supposed to be God in human form and the ultimate intercessor between humanity and the creator, followers cannot even find shelter in him.

Let's return to Adam and Eve. God created Adam and Eve out of the dirt, but then banished them from the garden and they had to reproduce to populate the earth in the same way the animals do — sexually. Adam and Eve sexually reproducing makes sense, but then in Leviticus 15, God is speaking against sexual relations. At the very least, the God character seems to have some problem with women's periods: "When a woman has her regular flow of blood, the impurity of her monthly period will last seven days, and anyone who touches her will be unclean till evening" (NIV, Lev. 15:19-20).

In Karen Armstrong's book, *A History of God,* she explains that the root cause for this commandment wasn't necessarily a disgust for the carnal nature of man. It was about preventing a husband and wife from becoming too familiar with each other (Armstrong 77). It was an attempt at creating conditions that prevented the couple from taking each other for granted. Yet, when looked at through our modern lens and the repulsion to carnal nature that religious leaders espouse, we can see why the first read sounds a bit more totalitarian. One reason this unclean vibe might ring true for us at first glance is that as the religion of Christianity underwent severe Westernization, the repulsion with carnality would have ramped up. As the cohesion in the group grew to a dominant Western religion, the idea of carnality being lowly or gross would have played on the individuals' foundation for cleanliness. This would not be in the sense of wanting to be physically clean, but rather wanting to be a clean person as seen through the eyes of the group. Not being seen as a filthy member is key to tribal survival. Even more specifically, Deuteronomy 14:21 exemplifies this societal norm where it talks about not eating anything that is dead:

> Do not eat anything you find already dead. You may give it to the foreigner residing in any of your towns, and they may eat it, or you may sell it to any other foreigner. But you are a people holy to the LORD your God. Do not cook a young goat in its mother's milk. (NIV, Deut. 14:21)

In response to this verse, one might argue God is trying to prevent people from infection or disease by avoiding potentially disease-ridden food. This makes sense as far as the moral foundations theory, as proposed by Jonathan Haidt, that groups select behavior that ensures sanitation (Haidt). Yet the verse goes on to say that foreigners can eat the meat or you can sell it. This betrays the sanitation reading of the verse by showing a lack of concern for outsiders and goes to show further group selection — not the message for all that some think the gospels are.

The reason these logical errors exist in the text is not only because the stories are typically allegorical, but also because they are created and documented by humans in a specific time and place and culture.

One of the fundamental reasons we must resist taking a hardline, literal approach to reading the

Bible or any ancient religious text or myth has to do with the core functions of our brain. In the book, *Proust and the Squid: The Story and Science of the Reading Brain,* author Maryanne Wolf describes some highly complex functions of the brain that must work in perfect unison for us to comprehend what we are reading. Wolf, who was at one point was the Professor of Citizenship and Public Service and Director of the Center for Reading and Language Research in the Eliot-Pearson Department of Child Study and Human Development at Tufts University, explains that while reading, you are engaging many cognitive systems such as attention, memory, visual, auditory, and of course linguistic functions (Wolf, 7). Concerning the linguistic function of the brain, research of the kind conducted by the psycho-linguist David Swinney shows that when you read a specific word, the brain calls up every potential meaning of that word and then applies the definition that is relevant to the context in which that word is occurring. The ability of the reader to accurately understand the meaning of what is being read depends on the paradigms available to the reader. These paradigms are, of course, shaped by lived experience. Wolf explains how children who have a diverse vocabulary will experience what they read in very different ways from children who have a limited vocabulary. A story that involves the main character going to a foreign land and experiencing the complex emotions of being in an unfamiliar physical place will conjure up very different paradigms in children who have traveled outside of their hometown, literally and figuratively, than it would in a child who has never left their small town either physically or in the form of mental stimulation. What we see here is that we humans can only grasp the meaning of what we read insofar as it relates to our lived experience and vocabulary.

Text is contextually limited by the time, culture, language, and personal experience of the reader. What does this mean for our reading of the Bible? The Bible was written and experienced by real people in vastly different times. The writers and listeners of the Bible had life experiences that we know very little about. The paradigms that are called to mind by early listeners might never match those called into our minds. Take, for example, one of the few references to Hell as an actual place in Dante's *Inferno*. Readers in the Medieval period took Dante's descriptions of hell as a writhing, glowing, carnal jail cell as literal and they subsequently feared it. Modern readers, however, now study the classic work as a piece of fiction that now provides insight into how the culture may have understood their biblical readings. That's not to say, though, that *Inferno* doesn't still influence many believers' perception of hell.

The word used frequently in the Bible for hell is *Gehenna,* which refers to a burning trash dump in Jerusalem. Very different imagery will arise in the mind of the ancient and the modern reader. This will, in turn, shape the way the different readers understand the passage. The reason the Bible happens to contradict itself, advocate violence, become hung up on blended fabrics and beards, and condemn the shedding of blood, is because human beings wrote it. These authors wrote their stories centuries apart and in different regions of the world, not to mention that they did not have the completed assembled books for reference before adding their own.

I have two favorite instances where the writer's voice in the Bible seems to truly shine through. One is an error when talking about insects, and the other is about poop. In one verse, the writer explains that God said followers should not eat four-legged insects such as locusts (NIV, Lev. 11:22-23). I have read defenses of this particular verse, such as can be found on the website for the Institute of Creation Research, that aims to show that our modern classification of insects, of course, did not apply at the time this verse would have been written. In his defense of the verse, John Morris explains that the front four legs of the locust are used for crawling while the back two are used for jumping, therefore the Biblical description is correct (Morris). The problem with this argument as a defense of the text is that even without modern-day classifications, any observer could see that the two legs used for jumping were still legs. Locusts still use them to walk, they are just the main ones used for jumping. Also, the fact that the defense is using the lack of modern-day classification betrays the fact that the writers were human beings capable of making mistakes. If an inerrant God was inspiring the text, he would know that locusts have six legs. The faithful prophet would have written "six-legged" because that's what God said even if the common assumption was four legs. The all-knowing, highest character has forgotten how many legs He happened to give those creatures. What seems more likely, however, is that the writer had a particular disgust when it came to insects and didn't know that locusts have six legs while he was trying to use God to outlaw their consumption.

My other favorite instance is when the writer claims that God has commanded that men in their camps should dig a hole and bury their solid waste because God moves among them in the camp and doesn't want to step in it.

> You shall have a place outside the camp, and you shall go out to it. And you shall have a trowel with your tools, and when you sit down outside, you shall dig a hole with it and turn back and cover up your excrement. Because the Lord your God walks in the midst of your camp, to deliver you and to give up your enemies before you, therefore your camp must be holy, so that he may not see

anything indecent among you and turn away from you. (English Standard Version, Deut. 23:12-14)

It would seem odd that a God that no man can look upon and that cannot even be in the presence of filth or sin would make the mistake of stepping in a pile of human poo. Also, the story of Genesis makes no mention of Adam and Eve not needing to have bowel movements before eating the fruit, so defecation would have been part of the original creation. God didn't have this law for the original human beings, and they didn't even know they were naked, never mind the fact that poop might be a problem. God walked among those two! Once again, what is obvious is the use of the God character and sanctity to create group cohesion by selecting for a behavior that favors sanitation.

When we consider one of the fundamental ideas of a creation where God gives free will to humans and wants, *hopes* that those humans will choose to follow Him to dwell with Him forever in the afterlife, we run into some serious contradictions. Once again, if we force a literal lens on the story of Eden from Genesis, the story will fall apart and, in this case, will be blatantly contradicted by later chapters of the same sacred text.

The many verses and chapters that seem to demonstrate a vicious and terrifying God, one that cuts down the enemies of His followers, isn't that surprising when you consider that it was written by people who were constantly in the midst of power struggles and regional violence. If the god of the Bible was less an omnipotent deity and more... Captain Hebrew who fought Romans and Caesar in papyrus comic strips, then it would make a lot more sense. People dedicated their lives to the gods of their lands. This God just so happened to be better than the *other* gods, better than Caesar and played the ultimate trump card by being able to do anything and cut down any enemy that came His way. It sounds like a superhero created to be better than your superhero, one created to calm the nerves and sadness of a population beset by violent occupation.

Instinct Versus Intent

How do we draw the line between instinct and intent? Instinct is generally defined as any fixed pattern of behavior in response to a clearly defined stimulus ("What is INSTINCT? What does INSTINCT mean? INSTINCT meaning, definition & explanation"). This means that a behavior that comes about without the basis of prior learning or experience is instinctual. For example, newly born sea turtles will instinctively make their way to the sea upon hatching. There are no adults around to tell them to make this journey, rather their instinct tells them this information.

In *The Soul of an Octopus,* author Sy Montgomery describes the job of Wilson Menashi in caring for an octopus. Menashi is tasked with creating complex toys for the octopus to play with and puzzles for it to solve. For the octopus, solving a puzzle results in being able to access the food inside, therefore receiving an award for solving the puzzle. The octopus interacts with the toy, determining what moves open which compartments or adjusting the puzzle to progress further. The octopus learns these traits of the toy or puzzle and the next time the same toy is introduced, the octopus reaches the solution much faster. Why? Because solving the puzzle one time and learning how to utilize various movements has helped the octopus with a new experience. The octopus adapts prior experience and applies it to a current one to become more efficient. This behavior could not be defined as instinctual (aside from perhaps the drive to acquire food) because it is learned, not innate. Numerous examples of other animals' behaviors that do not have the underlying base motivation such as food have been recorded. The fact that animals exhibit clear learning, intent, decision making, and execution of the decision shows self-awareness and free will.

The strategies used and time taken vary with the personality of the specific octopus, however. In these simple interactions, we can see two things. First, we see the ability of the octopus to maintain its focus on a goal and use critical thinking and fine motor control to problem-solve. While the drive for food is a basic instinct, this kind of learning and adaptation is not. Second, the rate at which the octopus can solve the problem is a function of its personality, meaning the information processing is not simply going through a basic instinctual drive for food, but rather is being filtered through a unique mental framework. The octopus is making conscious, calculated, problem-solving decisions that it does not have to make. It's choosing to engage in this exercise for the reward of food. In such a simple interaction, we can see, on a local level or individual-specific context, free will and intent at work.

This understanding of free will and intent must include behavior known as reflexive. Reflexive responses are involuntary and can include (but are not limited to) instinctual behavior in response to stimuli. In these moments of reflex, an animal's instinct-driven behavior becomes the default response, such as a dog barking at a noise outside. It is at this point that a special part of the brain called the lateral frontal pole prefrontal cortex becomes the dividing line. In creatures — such as ourselves — that have this specialized part of the ventrolateral frontal cortex, there exists the ability to interrupt or slow down reflexive behavior and weigh the cost and benefit of the decision ("Brain Area Unique to Humans Linked to Cognitive Powers").

In creatures where this part of the prefrontal cortex is not as developed, the instinctual behavior is more likely to be acted upon without any thought process behind it. It's important to note, however, that this is not a black-and-white issue. Just because an animal with a lesser or absent lateral frontal pole prefrontal cortex responds reflexively based on instinctual behavior does not mean that an animal is devoid of self-awareness, conscious experience, or any other subjective feature of the mind. It is simply a different kind of experience.

Consider the classic mirror test, for example. The mirror test is a study that tries to determine if a creature has any amount of self-awareness. The researcher will secretly place some kind of paint or marker on an animal in a place not visible to the creature without aid, and then place a mirror in front of the animal. The key in this setup is the secrecy in placing the marker on the animal so that it doesn't have a predisposition to inspect that area. Once in front of the mirror, researchers will watch carefully to see if the animal sees the marker in the mirror, and then whether it tries to clean it off of its own body. This would suggest that the animal recognizes itself in the reflection and therefore has some sense of self and can differentiate itself from other beings.

This study has been done on numerous different species, such as elephants, apes, cats, dogs, fish, etc., and many of these animals show quite plainly that they do recognize their reflection as just that — a reflection of themselves. This study is helpful in our attempts to glimpse into the conscious experience of another creature because it indicates self-awareness, but it doesn't suggest a reason to anthropomorphize the creature's experience. For example, the mirror test was done with ants and it showed that individual ants do understand that they are something different from other things around them, but that's all we can glean from the experiment (Cammaerts, Marie-Claire, and Roger Cammaerts 521). It's a fault of ours to assume our specific kind of awareness is the stick by which all others must be measured.

When we consider the differences between ourselves and an octopus, a honeybee, or any other creature, we can't help but feel superior. After all, there isn't a species that can outpace us on the food chain. We seem to have established dominance over any other creature we encounter. We've created the largest colonies that impose their wills on the landscape. Any circumstance that attempts to hinder our preference for easily moving through the world will soon meet its demise if not already even if at the expense of our environment. Doesn't all this prove that we are different from the animals? Doesn't it show that we are the more advanced, more successful, more evolutionarily superior creature? Didn't we "win" the game of survival of the fittest?

These are valid points and arguments that can be proven by examples constrained to this specific context of existence. However, all these points have their basis in the critical assumption that exerting our will on our environment is the measuring stick with which to determine advancement and superiority. This is a misguided assumption. We must start from a different perspective and ask the questions, "What does it mean to be evolutionarily advanced?" and "What does it mean to be a successful species?" The most common criterion we typically use is the brain and how developed it is. Our human brain is an incredibly costly evolutionary trait. It requires so much fuel and other bodily resources, yet it is arguably our greatest asset. With our highly developed lateral frontal pole prefrontal cortex, we can conduct introspection and abstract thought like considering our place in the universe.

Yet is having such a developed brain — or even a brain at all — a reasonable criterion for evolutionary fitness? Consider this: modern humans seem to have come about only in the last 200,000 years or so (Howell). While modern humans' actual starting date is continuously being found to be much earlier than we ever thought possible, as you will see when we consider Noah's story, civilization as we currently understand it is very recent, having come about only 10,000 years ago (and again, this timeline is fluid as more research and discoveries come to light every week). As far as the timeline of the Earth is concerned, humans, modern or not, have existed only in the very late chapters of the story.

A honeybee's brain is composed of fewer than a million neurons. Contrast this with a human's brain containing somewhere around 85 billion neurons (Chittka, Lars, and Niven 1). But don't let the honeybee's size fool you. This sesame-seed-sized organ is incredibly dense and complex. It allows the insect to use its highly advanced sight and smell to move through the world. A honeybee can also be taught new information and recall this data as shown by an Australian study in which honeybees were taught to recognize a certain color and use it as a guide to navigating a maze. This may seem like a meager existence when compared to what humans have accomplished, yet honeybees have their origins much earlier in the story. The oldest recorded honeybee fossil is estimated at around 74-96 million years old

(Michener, Grimaldi 1). It is estimated that the majority of the honeybee's evolution took place during the Cretaceous period. This tiny creature had existed long before humans came on the scene. All the while and since they have carried out their duties pollinating plants. Without the honeybee carrying out this task, humans may not exist, at least not in the way we do now. According to the American Beekeeping Federation,

> Many of the country's crops would not exist without the honey bee at bloom time. Crop yield and quality would be greatly reduced without honey bee pollination. As honey bees gather pollen and nectar for their survival, they pollinate crops such as apples, cranberries, melons, and broccoli. Some crops, including blueberries and cherries, are 90-percent dependent on honey bee pollination. One crop, almonds, depends entirely on the honey bee for pollination at bloom time. ("Pollination Facts: Honey Bees Are Pollinators")

It's estimated that one in every three bites of food we take is derived from plants pollinated by bees (Suzuki). This means that we absolutely must have bees for us to go on existing as we have thus far.

The inverse of this is not necessarily true. With their deeply ancient history and critical role in the ecosystems of almost every population on the planet, honeybees could thrive without human beings in the picture. Does this still make us the winners? Can we still claim that we are the most successful, most well-adapted, top-of-the-evolutionary-chain creatures we think we are? Our existence is dependent on many biological systems and relationships that have existed long before us, including one as small as a honeybee.

Let's take this a step deeper. Consider the jellyfish. There are not many fossils of jellyfish (as they have no bones) to get a clear understanding of their history. However, in an article from 2007 in *Science Daily,* the author explains how researchers found a "soft" fossil of a jellyfish that is dated to be more than 500 million years old (Public Library of Science). These creatures have been bobbing around in the sea for so much longer than humans have (or honeybees for that matter) that it's nearly impossible to wrap one's mind around that length of time. What might be most interesting of all is that jellyfish do not have a brain at all, but rather a loose neurological network. What's more, they don't have a heart, their bodies are almost completely water, and they can be highly venomous. Are we the winners of evolution? While we have exploited our abilities to the highest degree to assert our dominance in our surroundings, we are young pups compared to honeybees and jellyfish. Since they have been able to survive for so long, experiencing such dramatic changes to their environment and persist in harmony with their surroundings suggests that they are more adept in this world than we are. Again, we tout our achievements and advancements as proof that we are the highest-evolved species on the planet. After all, we do own dogs and cats, not the other way around.

Let's look at an example of a rather complex system specific to humans: government. We humans can discuss this topic with other like-minded humans even if we don't have intimate knowledge of the process. This is because we and the person we are discussing with are of the same type of consciousness. Would we be able to sit our dog down and explain the importance of government to them with any success? Of course not. Our words would not be understood by the dogs and dismissed in the mind unless what we're saying involves key words that they have learned. A dog simply cannot grasp the concept of government. Does the dog see us as a more advanced or evolutionarily higher being? What is most likely is that even though we may provide food and shelter and care for our dog, it probably doesn't see you as anything other than a pack leader. It can only understand you so far as the limits of its concepts and world. It can't put you in a box that it doesn't have. This means that, try as we might, we can never really step inside what it is like to be a dog, and the dog will never know what it's like to experience government.

Now let's flip the dynamic around and consider a creature that is above us just as we are "above" the dog. We might have words for an entity like this such as alien, angel, etc. The key here is that while we can entertain the thought of a creature that is above us on the evolution spectrum, we can only understand the parts of their experience that we have mental frameworks or language for. Let's assume this "higher" creature can experience time differently than we do, so as not to be constrained by past, present, and future. We cannot know, no matter how hard we try, what it is like to have this experience of reality. We can theorize, hypothesize, and conceptualize, but we simply do not have the mental categories to put this experience into. Furthermore, this entity may not be able to lower their conscious understanding to our level to explain this higher-level understanding of the world any more than we can discuss family dynamics with a honeybee.

To get a better idea of what normal honeybee behavior is like, I reached out to a couple of local beekeepers who were kind enough to take the time to respond to my questions. I asked what caused them to want to start keeping bees originally. One keeper named Ross told me that his interest in the benefits of having fresh, raw honey, along with all the interesting facts about bees drew him into the profession. I asked what his relationship with the bees was like, and he told me that it's a special relationship. He spends his time tending to the bees in the hives as well as rescuing bees that have moved into an unwelcome space. Note that he used the word "rescue." If anything is a sure sign of a bee-lover, it's that they don't exterminate bees that have moved into a humans' residence. He rescues them. Ross also mentioned that bees cannot survive more than a couple of days without their colony. The colony is a superorganism and Ross believes that while the bee is conscious of itself, its main concern is for the colony. Then he said something very telling: bees sacrifice themselves every day for the good of the colony.

To further illustrate the complexities of the question of consciousness in creatures, I'd like to quote a tweet from Richard Dawkins, ethnologist, evolutionary biologist, and author.

> Do dogs go to heaven? Only pets, not wild animals? If only humans go to Heaven, how far back? Homo erectus? Australopithecus? Only people who lived after Jesus? Not Abraham, Moses, or Elijah, then. Which 'you' will be in Heaven? Senile you with dementia? Midlife-crisis you? (Richard Dawkins)

I feel that Dawkins' words speak to a similar idea as the one I present here. How are we determining where to draw the line on who or what has a soul and thus has a concern about the afterlife? When we start to pay close attention to the lines we have drawn, it becomes clear that the answer to this question is not as clear as we thought. After all, if we drew a line to expel animals — meaning they have no afterlife concerns — then it wouldn't be breaking the rules for God to step in and stop an Africanized bee from killing a toddler or someone deathly allergic to bee stings, yet the bees can and will given the right circumstances with seemingly no divine intervention. Free will in this context suggests that a creature who has an afterlife would be entirely independent. Returning to the fundamental argument of this book and assuming a literal reading of the Bible, free will implies a soul, a soul implies the afterlife, and the afterlife implies that God is concerned with our choices that affect that afterlife. We have seen that bees (and other animals, for that matter) have free will. They are making decisions that far surpass mere instinctual response, therefore they must have independence. Yet God *doesn't* intervene in their decision-making even though there is supposedly no afterlife for the bee. It seems that in our attempt to decide which animals go to heaven, we have fallen into the trap to anthropomorphize the world around us. Just like we anthropomorphize God.

These compelling examples of all kinds of creatures having a level of consciousness that allows for what seems like a certain level of free will have troubling consequences for the relationship between the creatures and God. To our earlier point, the reason God doesn't seem to be concerned with the actions that animals take is because they do not have a soul and therefore do not have afterlife concerns. However, since beings like honeybees and octopi mirror many traits with beings like highly conscious humans, can we be so sure they don't have a soul? Moreover, can we be certain that having a soul and having free will are so intimately linked? One potential takeaway from the creation story is that having the breath of life goes hand-in-hand with having free will.

If we look to the Bible for answers, things get even more confusing because of the way God chooses to use insects and mammals as tools, or more accurately, as weapons to further his goals. There are so many verses where God sends insects to destroy crops or wild animals to kill people, or to kill children. Recall our earlier question regarding the honeybees. Why doesn't God override the actions of a creature that doesn't have free will and prevent them from harming an innocent human or child? Well, it would seem that question misses the larger point: God will outright use these insects and wild animals for the direct purpose of harm. So even if all non-human creatures do not have a soul and thereby do not have free will, the God of the Bible, the God of honeybees, is still using them to bring about calamity. However, if they do have free will, God is violating the fundamental concept that he will not override a being's free will.

The mere existence of a creature that might bring about harm to a human being throws a bit of a wrench in the works when it comes to the way we understand this god. We find ourselves stuck between a rock and a hard place. Either God is bringing about violence using things He created, or He is breaking the foundation of what draws many to this faith: God's love is a gift to you that you may accept or reject. He gave you free will to make that decision. He isn't forcing you to love him. If you are a follower of this faith, does this sound familiar? Accurate? What are we to do then, if God is going to override a being's free will?

In Joshua 2, verses 16-20, God actively "hardens the hearts" of an entire population to coerce them to go to battle against Israel so that they will be defeated and Israel's people can move in (NIV, Josh. 11:20). Now we have gone well past arguments over whether or not honeybees have the same free will that we have. God has just overridden what supposedly makes humans human. These inconsistencies are more than simple oversight. They have very troubling consequences for humans under the omnipotent control of God. The creator will not step in to prevent the supposedly soulless honeybee from attacking an innocent victim, but He will overrule the hearts of humans to send them to the slaughter? Perhaps God didn't have the same kind of consideration for those humans because they didn't have souls like the Israelites did.

We can see that our biggest achievements and assets — a large brain, enormous colonies, and a dominant spot on the food chain — do not work as criteria with which we can determine our success. Success might more simply mean surviving or reproducing, in which case all living, reproducing organisms are successful. We can also understand the limits of our ability to communicate with a lower-level consciousness or understand a higher one. With that being said, who's to say these creatures around us aren't above us on the evolution chain? We wouldn't know if any of their experiences were something we did not have a category or word for. Could we then, in our limited consciousness, ever really know God fully? We have the capacity to relate to the divine so far as the human experience is concerned, but our understanding stops there.

We can see, even with the briefest of looks at animal behavior, that the line between the consciousness they embody and the one we humans embody is a blurred one. That line is not the only distinction that dissolves upon closer inspection. The framework we have around what our consciousness is isn't even well defined and has been argued over for about as long as we have record on human history. In the book, *The Origin of Consciousness in the Breakdown of the Bicameral Mind*, author Julian Jaynes tackles whether or not learning is the hallmark of consciousness. Jaynes explores how he tried a test with his plant to see if he could get it to start to associate a tactile input stimulus with a frequency of light by drooping of one of its leaves as a response. Julian concluded that the plant was not conscious as it could not learn to droop the leaf at the signal of the light. After related experiments with ever more complex organisms and yielding the same results, Jaynes finally concluded that using the ability to learn in this way was by no means a measurement of consciousness. He writes, "Why then did so many worthies in the lists of science equate consciousness and learning? And why had I been so lame of mind as to follow them?" (Jaynes 12-13). He argues that when we introspect, acting within consciousness, it is not dependent on some conditioned learning such as mazes or bells. This fundamental error in conflation of consciousness and learning, known as experience, has become embedded in the discussion. Jaynes provides an example later in the chapter regarding a lab experiment.

In the experiment, volunteers would have air puffed into their eyes, causing them to blink, right after a light signal. Eventually, the volunteers' eyes would blink as soon as the light signal occurred even without the puff of air. Jaynes' point here is that the twitch of the eyelid in response to the light signal is, of course, learned behavior, but it isn't under conscious control. The participant isn't consciously controlling the movement of the eyelid; it's more like a learned reflex. At this point, Jaynes drops one of my favorite quotes: "It is now absolutely clear that in evolution the origin of learning and the origin of consciousness are two utterly separate problems" (Jaynes14). This would suggest that while creatures like bees, octopi, and even slime molds are capable of learning, that capacity alone does not clue us into their natures and therefore experience of consciousness.

In his book, *Intelligence in Nature: An Inquiry into Knowledge,* Jeremy Narby comes across this idea while interviewing Toshiyuki Nakagaki, a scientist and professor at the Research Institute for Electronic Science at Hokkaido University. While discussing the capacity for learning in true slime molds, the scientist says that what is clear is that the slime mold is capable of computations as evident by its ability to accurately solve mazes and obtain and absorb food in the most efficient manner possible. He goes on to point out that this kind of learning relates to the creature's unconscious behavior. The analogy the scientist presents is riding a bike. While riding, the rider is not conscious of the position and tilt of every part of their body, rather the body learns and adapts in a more unconscious but no less effective manner (Narby 105). This is the same idea we have examined in Julian Jaynes work.

One of the many fantastic analogies used in Julian Jaynes book is that of a map or diagram in terms of the idea that consciousness can be isolated within a certain structure in the brain. Studies around the reticular activating system in the brain show that if it is severed or damaged, the patient will end up in a permanent coma. Alternatively, if that same region is electrically stimulated in a sleeping animal, it will wake up (Jaynes 24). While these results tempt us to attribute consciousness to the reticular activating system in the brain, we must resist that temptation. Jaynes points out that even if we had a detailed map or diagram of every single nerve ending and all of the connections they hold across the body, we still could not derive from that map whether or not that creature has conscious awareness of itself. What's more is this argument does not account for the countless reports of conscious experience while the body is in a coma state, be it from injury or anesthesia. All of this makes two points very clear. First, we share much more in common with the world around us in the context of consciousness than we may realize initially. Second, we aren't even sure of the nature of that which we claim sets us apart from all other life on the planet. Again, upon closely inspecting fundamental beliefs about how the world works, we can see we don't understand as much as we think we do. We see that in two realms we often take for granted, religion and our place among other life forms on the planet, we are still learning. There is still much to learn.

Relating to the argument I have presented here about other forms of understanding, the human experience of consciousness is also limited by form. Our human form and channels on which we might pick up the consciousness frequency are limited by our brain. To "know" something is to draw lines around it. Knowing something means we have delineated this concept from others, placed it in on a shelf in our limited mental bookshelf, and we may be able to recall it at a later time. Seeing the situation from this perspective shines a light on another dimension. To "know" something implies that we are outside of it. It suggests that which we "know" is an entity or form in its own right, and we are something else that has observed, classified, and categorized that thing.

It seems to me that this may be one of the driving forces behind humanity's never-ending urge to put form around its experience of the divine. We are trying desperately to draw lines around something so that we can "know" it and feel as though we have a firm grasp on how the world works so that we don't have to be scared anymore. Unfortunately, the form that is draped on the divine takes different shapes and colors depending on which group of people you ask. Early Jewish teachers understood this in their idea that God was not the same for everybody (Armstrong 74). The mistake we make is assuming that the costume we put on God somehow exists in its own right. We forget that we tailored that costume to match the party we were at, the group we hung out with. European and Western Christians transformed an average middle eastern man into a light-skinned, blonde-haired, pristine figure. The idea that we must wake up to is that we have tailored our costume as well. That which our gown is draped on is that which God is draped on. We don't need to pass judgment on ourselves for playing this game; it is after all the grand performance. However, if we wish to step out of the rushing water, we must understand this. Perhaps we should be more hesitant to assert one interpretation of the nature of God or God's words over another when our experience of this divine is limited to the human frame.

What About the Honeybees?

Let's discuss one creature in particular that has fascinated me for some time now: the Africanized Bee.

The Africanized Bee has a very interesting origin story. They were created by breeding African bees with a species of European bees. In the early 1950s, these bees were brought to Brazil in an attempt to help increase honey production. Unfortunately, not long after their introduction to South America, multiple swarms escaped captivity and entered the wild.

The bees then made their way across South America and eventually showed up in Texas. What makes Africanized Bees so interesting is not only their origin story but also their behavior. An episode of *Outside Podcast* called "Bee Still My Heart," describes a fictional interaction with a hiker and these bees and what it would be like to encounter a nest and be targeted by the swarm ("Bee Still My Heart"). You would be attacked from all sides with waves of militant bugs ready to sacrifice their lives for the sake of the hive. You would try to run as fast as you could from the swarm, but they would keep pace with you, the whole time sinking their stingers anywhere they could. At one point in the episode, the narrator describes how you might jump into a nearby river to escape the swarm, but the swarm would hover over the water waiting for you to come up. When you do break the surface, gasping for air, the bees would descend upon you, entering your mouth even at the risk that they will be crushed. The operating principle is that they will be able to sting you enough to perhaps bring you down even at the expense of their own lives. Of course, a swarm attacking an individual perceived to be a threat is fairly common, however, this is where one aspect of the bees' behavior stands out. Africanized Bees have been recorded pursuing targeted humans for at least a quarter of a mile.

Isn't that incredible? Imagine running as fast as you can for just under three-and-a-half football fields and these little bees keep pace the whole way. There's more. The bees that are chasing you have been reported to have killed around 1,000 people, and they deliver close to 10 times more stings than your usual bee ("Africanized Bees").

Intense.

Now we must examine this exact intensity. Do the bees' persistence and determination show a clear choice and intention on behalf of the Africanized Bee? Is it true that the bee is acting out of defense of its hive and not just out to kill? However, if it were merely an instinctual reaction to ward off a threat from the hive, why would they actively hunt you for 10 hours? And I use the word "hunt" purposely. Africanized Bees track you for hours and hours for almost a mile. That is choice and planning. If you stumble across a female dog with her pups and she decides to ward you off, she may bite and bark or even chase you, but it is unlikely that she will pursue you for hours and for that far. Defense against threats relative to their proximity would seem instinctual. Defense against threats as targeted individuals that remain targets regardless of proximity is choice.

I decided to look into bee behavior a bit more. I came across Thomas Seeley's award-winning book *The Wisdom of the Hive.* Seeley is one of the leading biologists in bee and swarm behavior who has even had a species of bee named after him. In his book, I found an immense amount of fascinating details regarding bee behavior. What caught my attention the most were the details that further displayed purposeful intention on behalf of the bee. Seeley points out that a colony of bees functions within a highly complex social framework. He also points out some incredible behavior of the colony. Seeley writes that a bee colony has some amazing characteristics:

> [A bee colony] monitors the flower patches in the countryside surrounding its hive; it distributes its foraging activity among these patches so that nectar and pollen are collected efficiently, in sufficient quantity, and in the nutritionally correct mix; and it properly apportions the food it gathers between

> present consumption and storage for future needs. In addition, a colony precisely controls its building of beeswax combs for honey storage, strictly limiting this costly process to times of clear need. And it adaptively adjusts its water collection in accordance with its need for water to cool the hive and feed the brood. (Seeley 6).

If we think again about the dog guarding her puppies, we might argue that her behavior is instinctual — protecting offspring from danger. Consider more of that dog's behavior. When she is fed, do you suspect that she would ration and measure how much of her dog food she would eat at each meal? Would she reject the dog food if it wasn't the appropriate nutrient mixture for her and the puppies? No, we have not seen a dog display such forethought as this. And yet, this is exactly what the professional researcher Thomas Seeley documents in the behavior of bees.

Jeremy Narby writes in his *Intelligence in Nature* about a researcher, Martin Giurfa, that showed bees moving from one plant to the next, yet only within the same species of plant. With brains the size of pinheads, a bee can be taught to connect what it learns through the sense of smell to what it is seeing through its compound eye. While Giurfa is cautious to point out that we ought not to take his study's conclusions too far (Narby 59), we can be sure of one thing: bees do have their own conscious experience of their lives. They aren't simply instinctual machines. They must abstract, communicate, plan, and sometimes strategically attack a perceived target. All of these actions indicate a choice that is far beyond reactivity or reflex.

Where do we stop chalking these habits up to instinct and start attributing them to conscious intention? In the book *The Soul of an Octopus* naturalist Sy Montgomery explains that discussing the mind — not the brain — of animals in a research context is typically seen as scientific heresy (Montgomery 7). The aversion to studying the mind instead of the brain seems to be because scientists are well aware of how animals differ from us. What's more, is that we have countless cases of fellow humans pushing the envelope of anthropomorphizing wild animals and meeting a terrible fate. Consider the Grizzly Man. Hence the reluctance to draw too many connections between ourselves and animals is an understandable one. I would argue, though, that this conclusion is a naive one at best. While it's true that mistaking a wild animal for a kindred soul can result in horrific consequences, it does not mean that there is no similarity between us and the animal. What fails us in drawing these connections is the assumption that the social and cultural constructs that we have created are mirrored in other creatures. If we drop this fatal notion, what we are left with is a line or connection between our human consciousness and the consciousness we find in another creature. The same animating energy that is our consciousness exists within everything else. It is simply filtered through a different lens of sensory input, communication, and group norms. While the honeybee or octopi have a different conscious experience from us, it doesn't indicate they have a lack of said experience or free will. If they have free will, do they have a soul and thus concerns about the afterlife? However, accepting that creatures such as honeybees or octopi have souls raises some questions about what that means for the logic behind the afterlife presented in the Bible.

What does this mean for the dichotomy presented earlier? Well, assuming the biblical logic that free will and having a soul are intimately intertwined as they relate to the afterlife, I would say we have to come down on the side that creatures *also* have souls. God created man in perfection and a perfect garden. He gave the power of free will to humans for them to willingly choose His love. He is concerned with whether or not we choose His love because we as humans have souls that will spend eternity either in Heaven with God or separated from Him.

Based on this logic, it would seem that the only beings in creation that have any plan for the afterlife would be those with souls since the soul is what will spend eternity with or without God.

Would it be fair to also say that only beings with free will would spend eternity with or without God? Given the logic we just presented, having a soul seems predicated or perhaps intertwined with having free will.

A common practice in many Christian doctrines is that animals simply cease to exist after death as they do not have souls. I learned from a Catholic friend of mine that this view is held even within the Catholic tradition although it does undergo different interpretations depending on the Pope.

Let's assume just for a moment that animals *do* have souls. This would mean of course that they would be under the same framework we described earlier, namely that they have the free will to choose or reject God and this will determine their eternal fate. The concept that animals have free will would have serious implications for the next time one's pet chooses to defecate in the house. Also, the Bible does not suggest saving the souls of animals and, as far as we know, there was no Jesus-dog or Jesus-cat. It has been argued, though, that animals already know God. Yet, how would one know God and supposedly have reverence for him while not having a soul, thereby not having afterlife concerns? The creature must then have a soul like we are assuming. But then, why no mention of this in the Bible?

Let's now assume that animals do *not* have souls and therefore have no concern for the afterlife.

This would mean that they might not have free will to choose or reject the Creator in the way that humans do. Following the reasoning presented, this would mean that God is not concerned with the choices that animals make while alive on Earth.

In *Intelligence in Nature,* Narby suspects the reason the term "intelligence," when attributed to plants or nature as a whole, makes Westerners at best skeptical, at worst indignant, is the influence of Christianity. He says this is because Christianity makes people assume humans are the only ones with intelligence (Narby 99). Therefore Western researchers do not look at natural specimens as if they have intelligence, so they do not find intelligence. This is exactly what I am trying to get at here. Modern research is opening to the reality that intelligence is all around us. We are not so unique in this regard. What does this mean for a literal Christian outlook on life? What happens when we find out that even the honeybees have intelligence? And for how long will we continue to assume we are the sole bearers of a soul?

Given our look into what such tiny creatures as the honeybee is capable of, it must be said they display very real free will within the context of their specific existence. Based on the origin of free will presented by the Bible, it directly correlates with the entity having a soul. Therefore, that soul-bearing entity, within the afterlife image given by the Bible, would have afterlife concerns and God would need to be concerned with that entity's actions on this earth. The fundamental concept of humans having souls and free will links those two aspects together. Yet, we can see that other creatures display remarkable capabilities that are clear examples of free will. And yet, the Bible makes no mention of the animals having eternal souls. We have shown the relationship between free will and having a soul. We have also shown that other creatures display free will. At this point, it seems either the Bible is wrong for not including other creatures in the soul-bearing category, or the framework of having a soul and the afterlife is faulty.

What Falling Asleep and Spilled Pills Told Me About God

God topples from the sky, hell's fires fade:

Exit seraphim and Satan's men:

I shut my eyes and all the world drops dead.

I fancied you'd return the way you said,

But I grow old and I forget your name.

(I think I made you up inside my head.)

I should have loved a thunderbird instead;

At least when spring comes they roar back again.

I shut my eyes and all the world drops dead.

(I think I made you up inside my head.)"

Excerpt from Mad Girl's Love Song

By Sylvia Plath

I introduced this chapter with an excerpt from this lovely poem by Sylvia Plath because I feel that it speaks to the deconstruction in my own life. The poem gives voice to the tendency for humans to make up our own God in our minds. While there have been many events that led me to my current view, the most recent happened while I was at work. My most recent job had been driving a box truck full of milk to grocery stores. To make the deliveries in time, I had to get up very early — 1:30 AM early — every day. Once I loaded the truck, I was off to what was usually a fairly long drive. When I first started, I would end up driving at least an hour first thing in the morning. Here's the thing: I hate driving. Not because I get bored or stressed, but because I can't stay awake. This made my morning trips pretty terrifying for me. I would come to and not know where I was or what exit I got off of.

I remember one particular morning when I had a long drive ahead of me. I had loaded up and had to drive about an hour away. It was early and I was extremely tired. Regardless of how long I had slept the night before, I was always tired. Nonetheless, I climbed up into the truck and headed out. It was pitch-black and freezing outside. As I rolled along the monotonous road, the cab became cozy warm and the rumble of the truck rocked my seat side-to-side.

I hadn't been on the interstate very long before I started to notice the usual signs. My body would twitch all of a sudden. It felt like I had been zapped while having the sensation of falling. I'd reach for the radio or roll the window down a bit even though it was incredibly cold outside. I'd tell myself that I was okay, and then after a few minutes, my hand would just fall lifeless from the steering wheel. I would regain control of my arm and hand just before it landed in my lap, and my heart would begin to race.

It was at this point that I began to pray.

I asked God to please just help me wake up. Resituating in my seat and turning up the radio, I would feel okay for a second. Then my wheels would cross into the other lane or I would narrowly dodge a construction cone. And again, I'd snap back with full consciousness and heart pounding.

I would become frustrated. "Please, God, just help me wake up!" This cycle would grow worse and worse as it spiraled ever downward. Eventually, I found myself pinching my thigh and purposely screaming as loud as I could to wake myself up, only to doze off mid-yell.

I wasn't asking for much or anything selfish. I was only asking to be able to stay awake for my safety and the safety of others. Yet I received no help. What was it? Had I done something to offend God that He would abandon me now? What could I have done that was so serious that it justified this? If I have free will to choose or reject God, I was quite clearly choosing God and begging for Him to intervene at that moment. I was making the clear choice in favor of God, but my head kept nodding and my 15-ton truck kept changing lanes. I eventually made it to my first stop, but I couldn't remember much of the trip there or what may have happened due to my drifting truck.

Months went by before I was finally approved for a sleep study at the local hospital. The results weren't conclusive, so they prescribed me Modafinil to take in the mornings. Modafinil is not a stimulant, but it is supposed to help you feel well-rested so that you won't doze off during the day.

Aside from the nature of my job being a constant stressor for me, finances were also a problem. The money only got tighter once I had to start getting the Modafinil as my insurance coverage didn't help too much to cover the cost of the script. I would have to choose between groceries and the medicine that kept me from crashing. I was barely scraping by and struggled to keep my job. My boss would quite regularly remind me that if he had to let anyone go, it would be me since the sleep study was putting a strain on my qualifications.

One morning I was in the bathroom getting ready and I finished brushing my teeth. Then I opened the pill bottle to take my morning dose. As I tipped my head back to swallow the pill, something — I didn't see what — caused the pill bottle to tip over, spilling my excruciatingly expensive pills into the still-wet sink and then dissolve. Any chance I had of staying awake and ensuring I didn't cause an accident just melted down the drain along with the money I would rather have spent on groceries.

My inability to stay awake and the unfortunate loss of the pills caused me to consider what I believed about the nature of God and His interactions with us. If I ever managed to wake up enough to navigate the vehicle I was in or happened to *not* go in the red, finances-wise, I would thank God. I found, however, that when these things didn't happen, I would become so blindingly angry with Him. This is perhaps a common experience for many believers.

I realized that if I'm going to attribute the good things that happen in my life — such as managing to wake myself up and not kill anyone on the road — to God, then the lack of intervention on God's part must also be attributed to Him. Essentially if God gets the credit for the good, He must also take the credit for the bad. These particular scenarios are important to me and the idea in this book because they relate to the idea of free will.

For example, consider me falling asleep while driving despite my desperate efforts to stay awake. In that moment, I'm not choosing or willing myself to sleep. My body is reacting in a way that I obviously cannot control. This means that if God were to step in and wake me up it wouldn't be a violation of my free will. I was begging, *yelling* for God to intervene with something as simple as that.

The same goes for the pills, right? I didn't choose to dump them, and God knows I desperately needed them. I know that an argument will be made that I can't know God's will but that I just have to trust it. I understand this point, but should

I have said that at the funeral of someone I killed with my 30,000-pound road missile as I dozed off and drifted into oncoming traffic? Should I say that God could have helped me wake up, but He must have had some other plan that included the death of your loved one? I solidly disagree.

All that being said, I still want to make clear that the conclusion I came to as a result of this view of God did not result in me throwing out the idea of God all together. My situation with falling asleep and the pills are not an argument for why God doesn't exist. It is, however, an argument to call into question what we believe about the nature of God. As we discussed earlier, in a literal interpretation of the Bible, God created the choice between Him and His antithesis. Also, if He chooses to intervene, then He also chooses *not* to intervene. This doesn't mean there is no God, but it means that He is both sides of the coin, the right hand, and the left. Understanding that the dichotomy between Good and Evil is a false one can expand our concept of God. For me, this understanding has shifted from what I once saw as a good God with an army of Angels at war with the Devil and Demons, to energy or consciousness that is the foundation of not only our being but the being of everything. It is more like a framework that gives rise to existence, not a deity that is only concerned with our afterlife.

Peter Abelard, the Medieval scholastic theologian, claims that the root of good and evil lies within our intentions rather than our actions or desires (Abelard, Peter, and Paul Vincent). According to Father Kerry Walters, award-winning author of 40 books and hundreds of articles, Abelard argues that the way we know an intention is good or bad is whether it is in accordance with or in rebellion against the will of God (Walters). Abelard writes that our intentions are good if they conform to what our conscience and reason tells us is the will of God (Walters). This has particular relevance to the scenario I described in which I was drifting into other lanes. I intended to stay awake and not cause an accident. Was this intention in accordance with the will of God? I don't have a way to know for sure, but I would feel comfortable assuming God's will did not involve me creating a potentially deadly crash. So if my intention was good, as it is not selfish to pray that I do not accidentally cause harm to another, and it seems safe to assume God would not want me to hurt anyone, and my sleepiness was not of my own free will, then what was the delay on God's part?

I came across a video of Ryan Reeves at the Historical Theology at Gordon-Conwell Theological Seminary explaining the importance of William of Ockham after which the famous concept of Ockham's Razor is named. Reeves explained how Ockham poked holes in the scholastic arguments from some of the most prominent thinkers of his time as well as the scholastic period, such as Thomas Aquinas. Ockham was arguing that the ideas that the scholastics were promoting were built on a fabricated foundation. They had essentially made assumptions at the beginning of their argument that allowed for favorable outcomes at the end (Reeves). One of the fundamental "assumptions" is in Aristotle's Categories. A discussion about what the 10 categories mean and how they are applied to the world is a very dense subject, as well as highly debated by individuals with more substantial credentials than I will ever acquire. Suffice it to say that the theologians Ockham was arguing against have essentially taken what seems to be a simple classification system (the 10 categories) and turned them into entities of their own. They would then build from these assumptions' doctrinal ideas. What Ockham aimed to do was show how the scholars had simply adopted concepts that allowed them to connect certain ideas.

I think that Abelard did the same thing that the other scholastics of his time had done; they built their arguments on a fabricated foundation or assumptions that allowed them to keep going with their argument. While Abelard's ideas of using reason and logic to determine the morality of one's intentions make a lot of sense, it is predicated on the idea that the Bible alone has the last word on morality and it is against this measuring stick that we should judge one's intentions. One of Ockham's most pointed arguments, according to Reeves, is that the error scholastics made was assuming things existed that they could not prove and then building frameworks around that. I think Ockham's Razor did not go deep enough. I think that his thinning of theological thought is still based on convoluted and self-contradicting ideas about the divine.

The Razor is helpful in that it guides us to simplify our approach to any given concept. This is why I have aimed for simplicity in this book. Simple turnings of the gears of logic or understanding the context of certain ideas stop the storm of argument before it has a chance to begin. We need not wade into more complex thoughts. Through closely examining the story of Adam and Eve and the contradictions we see in the fundamental concepts presented by the Bible against what plays out in the natural world, not to mention the contradictions and inconsistencies in the Bible itself, how could one possibly use it as the metric by which to judge one's intentions? To borrow an argument from Ockham, claiming the Bible to be the foundation is to undercut all further ideas that stem from this foundation. Let's take it a step further. I would argue that there are entire categories of philosophical thought that are predicated on a shaky origin at best.

Prayer

Prayer is an aspect of belief that has affected me for most of my life. It could be credited with being the catalyst for my deconstruction and examination of the belief structure I had held all my life. To examine prayer closely, we have to first start with the question, "What do we hope to accomplish?" Secondly, we must find what underlying assumptions we are operating with. Let's start with the question regarding what we aim to accomplish through prayer. For this, I'll use a personal example, as trite as it may be.

I had been making deliveries all over the city. One day I left my last stop and began the long trek back to the warehouse where I would end my day. When I arrived, I walked inside, placed my stuff on a desk, and began my end-of-day process. When I had completed all I had to do and began to leave, I noticed that I couldn't find a small, handheld printer that I was carrying with me earlier that day. As I started to walk back inside, I said a little prayer, asking, "Please let that printer be inside somewhere." I asked this because if it wasn't inside, I would have to drive all the way back to my last stop to find it, and I simply didn't want to do that. Trivial, I know, but this is an example of the tiny prayers that we make daily.

It was at this point that I realized something: this prayer was pointless. It wasn't because I thought it was too trivial or a realization born from an atheistic perspective but a logical flow from within a theistic world view.

First and foremost, I realized that at the exact moment when I was saying my prayer, a certain reality already existed and certain decisions already made, certain actions already taken. At that moment of my prayer, I had already placed the printer down somewhere — either at my last stop or with the rest of my belongings that I had carried inside.

Next, it became apparent that I was asking God to do one of two things. If I had made the necessary decisions and taken the steps to ensure that my printer was with me, then I was asking God to allow that to be the reality. Alternatively, if I had *not* made those decisions or taken those steps, I was asking God to change the course of history so that my decisions in the past would result in the version of reality that I most preferred. This is what I hoped to accomplish with this prayer. Before we move to the next point, it's important to point out that even without further examination, this prayer is pointless even at this stage. If I took the actions necessary for my preferred reality, then God would have nothing to do with that reality being true. If I didn't, then God is still not in the business of changing history (as evidenced by history books) and my prayer is therefore pointless.

From there, I noticed that my prayer rested on one of two assumptions: either God was an individual entity that can choose to get involved in my life (or not) and had a distinct plan, or there was no set plan. If it is the case that God has a specific plan for my life, then why should I be praying for a specific outcome? If God is to be trusted with the responsibility of orchestrating our lives, should we not have such trust as to not utter a sound? If one could change God's plan merely by request, then the sanctity of the plan is not as sound as we thought. Alternatively, if God does not have a predestined plan for us, then why am I asking him for a specific result? If He is flying by the seat of His pants as much as I am, then He should not be looked to as the ultimate authority on the course of our lives. This unsettling notion is addressed directly by the famous author and social critic, Christopher Hitchens when he says "The man who prays is the one who thinks that god has arranged matters all wrong, but who also thinks that he can instruct god how to put them right" (Hitchens). This should be the most alarming news about God thus far since that would mean that He answers certain prayers but not others that perhaps do not adhere to His plan. So if we aren't praying to a separate entity, who might we be reaching out to?

I suspect this habitual urge to reach out to some power larger than ourselves might have its roots in humanity's affinity for creating entire myths, stories, and legends to put words to its experience of the divine. Perhaps the urge is not distinctly religious for some people as it was in me — maybe they take a simpler route by hoping for a specific outcome or reality. While some may argue, including myself, that the religious aspect of prayer is misguided, I do not think it is devoid of all-purpose. We will get into studies relating to this in a later chapter, but for now, we are simply examining the logical errors that prayer falls into. What is significant here is that we aren't necessarily arguing whether or not God is powerful enough to answer the prayer, so much as we are pointing out the logical steps the scenario must go through. Again, to bring up Ockham's Razor, we can't justifiably wade into discussions about what kind of prayers God does and doesn't answer, what he has the power to do, etc. because that entire discussion is based off a faulty assumption that the intention behind prayer isn't self-defeating.

This chapter is short and simple, but I feel like it must be. Simplicity is the name of the game through this entire book. A simple illustration to describe local-context free will, the idea of God chased away by honeybees, and here a simple examination on the goal of prayer. A simple, clear idea can be more effective than an entire book written on the subject. This is why I want to share these ideas with you; they are so simple. We don't have to wade into the more complex and murky topics if we start at the foundation and make sure it is sound. We must take back our agency and responsibility to examine principles and discern truth for ourselves. When we look closely at what informs our beliefs around prayers and the deity that hears them, it becomes clear that the underlying assumptions we have made are in exact opposition to each other. Simply the implications of a prayer being heard by the deity we are examining undercut the idea of that deity in the first place. And yet, we pray. Whether it be religious, reaching out to a god, or it amounts to not more than talking to ourselves, we make big and small requests of the universe constantly. As I said, the idea of prayer isn't void of all use. Many studies have shown a measurable effect of human intention on the physical world, as we will see in the next chapter. If it's true that we can bring about change in this way, what does that imply about the intimate connection we must have with our world?

Name the Space Behind Our Eyes

If I were to say to you that being obese will lead to increased blood pressure, impair your body's ability to modulate glucose, and cause elevated triglycerides thereby increasing your risk for heart attack and stroke, you'd probably agree. One reason you may agree is that there have been countless studies to support this conclusion. These studies and their conclusive results are the reason that our medical professionals treat us the way they do. That, and a consumer response indicating that we do not wish to be obese and that we as a society recognize it as a problem. One such study in the *Journal of Gerontology* showed that this relationship between the aforementioned symptoms and obesity had a P-value of .001 (Meyers). This means that there is less than a one-in-a-thousand chance of being wrong or put more plainly, there is less than 0.1 percent chance that being obese *won't* affect these physiological processes.

We should note here that the P-value is a calculated number, but it is used in unison with a pre-determined significance level or Alpha, along with sample size, so to use the P-value alone could be misleading. The significance level is greatly determined by the size of the sample and the number of tests. The number of examined individuals in the study I just mentioned was 132 otherwise healthy, non-smoking, adult men.

There are countless correlations and notions that we —as a society — hold true, and these widely-accepted notions actively influence our lives and the decisions we make daily. We are aware of these correlations as a result of studies concluding such a high probability of relation. For instance, this is how the effectiveness of all our medications is determined. We put quite a lot of trust in these studies (and not always consciously) every time we take an aspirin, a Tylenol, or an antihistamine. We assume that since it is an approved drug being sold at our pharmacy, it won't cause catastrophic changes or immense pain in our bodies. Further, we put trust in these medicines because the effect they have on our bodies, on our symptoms, is something we can prove. We have data. The clinical trials show that the effect of the drug is statistically high enough to prove it wasn't just chance.

It would seem that we can agree that studies with such a significant probability against chance should be taken seriously and have real implications for our lives.

Fair?

With all this in mind, let me explain the premise of another study. Here's the setup: On a table sits a box about the size of a suitcase that generates essentially white noise or random bits of data, resulting in 0s and 1s.

What's unique about this random number generator is that it *isn't* using a computer program to spit out these bits, but that it's using quantum tunneling. This means that in the box is a circuit that is pushing electrons against some sort of barrier. Some electrons — unpredictably which ones — will pass through the barrier and some won't, which is known as quantum tunneling. When there is a large number of electrons passing through this barrier, this results in 1s being generated. Likewise, when a smaller number of electrons pass through, a 0 shows up ("The Global Consciousness Project – Ph.D. Roger Nelson").

Study participants are generally asked to sit in front of this box, not connected to it in any way, and "intend" or mentally coerce the box to generate high or low numbers. Now, a bit about data.

If you were to make an either-or action, for example flipping a coin, the chances of any independent result occurring would be fifty percent every time. It will either be heads or tails. Chance, therefore, would be fifty percent and anything statistically significant would have to be above fifty percent by some degree. For example, in your dataset, if you flip tails seventy percent of the time out of one thousand flips.

When conducting a study and analyzing the data, one must establish the mean, or average. This average represents the normal or expected area for a data point to reside in. If a researcher were to ask you to predict the number of heads or tails in a set of 100 coin flips, you would likely get about 50% of them correct due to the somewhat 50/50 chance of heads vs tails. If you were to get slightly higher or lower than 50% correct, this could be due to chance. If the researcher finds that a data point falls within a range that indicates it could be due to chance, he must accept what is called a null hypothesis. The null hypothesis simply states that the pattern in the data is one of many possible variations resulting from chance (What is a Z score What is a p-value). The next step is to measure how far a data point is from that norm. This is called the Z score. The Z score represents how statistically significant a data point may be. Further, we need to establish what is called a critical value. Imagine a typical bell curve; a graph that starts low at one end rises like a wave to the top, then mirrors that trajectory back down making a shape of a bell. Critical values will commonly reside at the edges of that bell shape.

Let's go back to our coin flip analogy. If you guess slightly higher than 50% of the flips correctly, you will have a higher Z score. However, if that Z score does not cross a certain threshold, the critical value, then we cannot reject the null hypothesis. This means that we must say that your score that was better than 50% could still be due to chance. A common critical value for Z scores is 1.96 (What is a Z score). If you guessed so many coin flips correctly that it resulted in a Z score of 2, then the researcher must reject the null hypothesis and assert that this is unlikely to have been due to chance. As the Z score increases, the odds against chance increase, making it more and more unlikely that the data point is a result chance.

Set distances away from the average are measured in what is called sigma. This number is another way of measuring statistical significance. The gold standard is a threshold of six-sigma, indicating that the odds against chance are over a billion to one (Radin 2018). If an experiment reaches six-sigma then the treatment applied is confirmed so thoroughly that further testing is not required to prove an effect. With all these details in mind, let's go back to the experiment with the RNGs (random number generators).

Arguably one of the most famous research institutes using RNG's to study any relationship between individuals and the machines is the PEAR research lab. In 1997 an engineer at PEAR named Robert Jahn, along with his colleagues, published a review that analyzed 12 years of data involving over 100 volunteers. The conclusion was striking; the output by the RNG's seemed to favor the intention of the volunteer. That is, whenever the volunteer wished for the output of the RNG to have an unusually high number of 1's, that is what the output would likely be. The magnitude of the effect on an individual trial was modest, 1 bit out of every 10,000 being shifted, but when looked at overall the odds against chance were an astounding 35 trillion to one (Radin 2006). Studies such as this have continued and the data has been analyzed. It has been repeated hundreds of times with thorough data collection and watertight design. In light of this, the data has reached that golded six-sigma threshold.

Psychologist Roger Nelson has taken the study of RNG's to a new level with the Global Consciousness Project. This project involves RNG's set up all around the world, running 24/7. During times in between major world events, the data represents an established baseline, usually called a random walk. When a major world event occurs, the data before and after the event is then compared to the baseline. This experiment is different from just the individual type we looked at before. The Global Consciousness Project is examining the interconnectedness of our consciousness, as well as the effect of grouped, focused attention. Data has been collected since 1998 and has covered 500 world events. The results of the data have reached over seven sigma. This means not only does it surpass the confidence level to sufficiently prove the effect, but it also has odds against chance of over a trillion to one (Radin 2006).

Consider the faith we have in the drugs we discussed at the beginning of this chapter. One of the most common drugs is Aspirin. Perhaps you are familiar with the commercials advising you to take one aspirin a day for heart health. You may also have been prescribed by your doctor Aspirin or Acetaminophen because it is safer for your heart. A simple google search of Acetaminophen in relation to heart health returns nearly 7 million results with the first page containing details on this treatment from Tylenol's official website. Under the Health Conditions section there is a page about Aspirin therapy which states that "The American Heart Association recommends people at high risk of heart attack should take a daily low-dose of aspirin (if told by their healthcare provider)" (Tylenol). This is a common treatment, however, when we consult a meta analysis on studies examining this very prescription the results do not bear the same confidence that our doctors may have us believe.

According to a study published in the peer reviewed journal BMC Medicine, a meta analysis of 17,000 participants showed an effect size of .12 (Leucht et al. 3). The upper limit of a small effect size is .2. It should be pointed out that this figure, called Cohen's d, is not the same as statistical significance. Statistical significance is a measure of how likely the outcome was due to chance whereas an effect size is a measurement of how important the outcome is. A medium effect size begins at .5, making the results of the Asprin studies fairly unimportant. Let's compare this with the effect size of the studies examined in Dean Radin's meta analysis of RNG experiments. According to the data he provides, the mean effect size of individual experiments is above a 3 meaning they are more than 3 standard deviations above the control group (Radin and Nelson 8).

What this proves is a connection between each of us in our consciousness, whatever that may be. Perhaps more specifically, it proves there is a connection between that aspect of us and the physical world. If we do indeed have this rather intimate connection between ourselves and a simple random number generator, how much more significant must our connections be to one another? Luckily for us there is a study about that.

Dr. Dean Radin conducted studies on the connection between individuals. One such study had to do with an individuals ability to sense whether or not they were being stared at. Early experiments by other researchers were of a simplistic design, involving clickers and blindfolds, and this opened them up to criticism and a discounting of their findings. To avoid this, Radin set up his experiment in a very different way. Radin had person A, who would be stared at, placed inside an electromagnetically shielded, double-walled, steel structure. This would ensure that there was no possibility for communication between observer and observed. Twenty meters away sat person B who would observe person A through closed-circuit tv (Radin 2018). To measure if there was a change in person A's physiology, sensors were set up to measure their heart rate, blood pressure, and other vitals. The study involved 26 pairs of healthy adult volunteers. When the observer was experiencing positive emotions, it resulted in a detectable change in the vitals of the observed. The data had a P value of .006 when the observer was experiencing positive emotions, and .0009 when experiencing negative emotions. This, of course, indicates a probability far beyond chance (Radin & Schlitz 2005).

Recall what we discussed about prayer. In those moments when we are asking God to bring about the version of reality that we most prefer, we are willing that version of the future to come about. If you consider the vast amount of data showing the very real effect of human intention on reality, we are not communicating with an entity that is separate from us, choosing whether or not to get involved, but we are communicating with the consciousness, for lack of a better word, that pulses through every single bit of reality. To be fair, the data I have presented is addressing human intention on something as simple as a random number generator, up to affecting someones mood. This does not prove our ability to influence the course of human events through mere intention. However, it does highlight an interesting connection between prayer and the human mind.

Perhaps the name of God is ineffable not because of its holiness, but because it is preverbal. It is preverbal because what name do we give to the space behind our eyes? What is the name of the space in our consciousness in between falling asleep and dreaming? What word can we use to draw lines around pure conscious awareness? It is before words, before concept. I'm not referring to our brain, I'm referring to... well, I suppose it's hard to say what I'm referring to. When you sleep, your mind goes somewhere that is void of all things, but it's not blackness. It's nothingness.

Does this mean that we should be able to always get the version of reality that we want? Of course not; horrific tragedies occur frequently and plenty of distraught individuals are putting plenty of brainpower into wishing events will play out differently. We may be able to create different outcomes and realities just with our minds, but we also may not. Not getting what we want every time, however, doesn't disprove our capabilities to do so. Our modern understanding of consciousness is only beginning to emerge, and it is far too early to postulate on the reasons why some requests come to fruition and some do not.

Literal or Allegory?

If we are concerned with adhering to a literal reading of the Bible, we have to answer some serious questions that arise. To start, I want to ask this: if God inspired the writer's hand to such an extent that we can take a literal reading of the Bible, then why do multiple accounts of the same event differ from one another?

If we are going to white-knuckle our literal interpretation of the Bible and the idea of our moral compass being based on something as wishy-washy as allegory, we have to answer for something significant.

We love the parable of the Prodigal Son, don't we? It resonates with us; the way the father with open arms welcomes his son back home. This story about forgiveness and opening our awareness to the party that was around us all time, informs many of our moral compasses. The simple tale helps guide us in understanding what true forgiveness looks like.

We also love the parable of the woman who lost one coin and did not rest from flipping her house on its end until she found it. It's easy to identify with the coin as God does not rest until He finds us. Heart-warming, surely.

But I have a few questions. What was the Prodigal Son's name? Or his dad? What was the name of the woman who lost her coin? How long did she live? Did she beget anyone?

Again, we run into problems when we take a rigid, literal interpretation of the text. These questions also betray an underlying truth about how we are relating to the information in the Bible. It is obvious that these parables were never meant to be taken literally, but instead, they were to be used as a story or illustration to memorably convey a deeper truth. The fact that we use the parable as a moral guidepost shows we are using fiction literally — and that's okay. We love the way the story of the Prodigal Son illustrates God's love for us, but by asking what the son's name was or who the lady with the coin begat reminds us that we are okay with allegory. The point I'm trying to get to here is why are we okay with allegory when it's obvious, like in the parables, but not in something poetic like Genesis. How much are we okay with? The story of the woman searching for her lost coin is, of course, a tale to convey a bigger idea and this is okay. But what if the story of Noah is allegorical to illustrate a larger idea? What if the ten commandments written by the finger of God are a parable? Jonah and the whale?

Where then do we draw the line between what can be taken literally and what we must view through a lens of open-mindedness? How much are we comfortable interpreting and how much do we feel must be literally true, engraved in holy stone? I suppose I'm not so much trying to definitively say where that line ought to be, but I am saying that the fact that this could even be a question should be reason enough to loosen our grip on black-and-white absolutes. The mere existence of questions when it comes to interpretation is reason enough to slow down.

One more point about literal or allegorical interpretations I have is about Noah. Based on various sources, Noah would have been born somewhere around 3078 BC and the flood would have happened around 2497 BC. (Stewart). According to the author of the Noah story in the book of Genesis, the entire world was wiped out and the only remaining beings were Noah, his family, and the pairs of every animal around that he stowed away on the ark.

Here's the problem: If we white-knuckle this story to somehow be scientific fact and a literal, historical account of an event, we have to answer for a couple of things.

Even if one rejects the concept of carbon dating, we have a wide range of evidence that certain cultures existed long before the flood. We can see that these cultures continued through the time of the flood. For example, the native Aboriginal people of Australia have been shown to have a DNA lineage that puts them in Australia some 31,000 years ago (Jozuka). As is the way with this kind of data, this timeline changes very frequently. The same day I wrote the above line, I heard an episode of one of my favorite podcasts, *Mysterious Universe*, in which they explained some of the latest data suggesting not only that the out-of-Africa theory might not be the whole picture, but that the Homo Sapien timeline goes much further back than originally we thought ("21.24"). Some emerging evidence does not fit into the out-of-Africa theory. One study suggests two independent origins for early humans; one in Africa and one in Southeastern Asia (Carriers of mitochondrial DNA). Some contentious studies, such as one by Steven Strong, outright deny the out-of-Africa conclusion (Strong). Researchers have found pottery artifacts from study sites in Australia have dated more than 180,000 years old. The intriguing aspect of this research is that the combination of DNA lineage evidence along with carbon-dated artifacts both exist.

If the entire world was wiped out in Noah's flood, how do we account for the different ethnicities around the world? Many other cultures around the world have a history that incorporates a giant flood, indicating that they were around afterward to tell future generations. The stories in the Bible take place in Jerusalem and other locations in the Middle East, and if Noah and his family were the only people to survive the flood and they are all presumably of the same lineage, then how do we have such diversity in the world? How do we have traceable DNA lineage that is continuous through the time of the flood? Yes, different ethnicities assert traceable lineages to the four children of

Noah and their spouses, but can we really take those assertions literally? And don't even get me started on how *one* family repopulated the Earth. There is simply not enough time between now and the receding of the floodwaters for the genes of humans to develop into the number of distinct races on the planet.

Another question is do we as followers *need* the Bible to be literally true? Take, for example, the *Bhagavad Gita*. The *Bhagavad Gita* is a Hindu text that depicts the dialogue between Lord Krishna and Prince Arjuna (*The Bhagavad Gita*). Their conversation is about legions and armies that fight on a battlefield. Only the story is a proverbial one, and the actors in these armies represent human tendencies and the battlefield is our mind.

If one was to read this text through a literal lens, they would miss the point of the story altogether. When you understand that the story is just that — a story — you can understand the moral it's trying to convey. At its root, even the names of the characters included aren't important. The truth the text conveys could remain unaltered even if you changed Prince Arjuna and Lord Krishna's names.

The other side of this coin is what it would mean for our lives in modern times if the Bible is literally true. There are some incredibly disturbing commandments that God gave the early followers that would be at odds with how we conduct ourselves in modern times. In the same way that we draw a line at how much is too much allegory, we seem to also draw lines with what rules of the Bible are currently relevant and which ones are safe to let stay in the past. Where do we put that line and why? A conversation like this is, in fact, a very old discussion. It is what some of the earliest followers of Yahweh did when they gathered together. In the early days of Christianity, there was an understanding that God was not the same for everybody. Individuals experienced God in different ways, and this led to describing God to be "like" certain things. The conversation around how to think about and interact with the divine was a constant one, literally resulting in the Talmud (Armstrong 73).

Do you see?

The text here doesn't have to be literally true for it to have a profound impact on its followers' lives. The moral of the story still comes through loud and clear. As such, the Bible would have no different impact on one's life whether it was a literal historical account or simply a conglomeration of fictional stories and proverbs. What matters is what we get out of it. What we can see here is often our line in the sand over what needs to be literal and what can be allegorical, is a subjective decision. What's more, often that decision isn't taking into account the full ramifications of where we drew that line. As we have seen and will continue to see, this is not black and white. Simply the existence of these questions is reason enough to loosen our grip and reexamine our beliefs.

Offloading and the Moral Matrix

According to *The ABC's of Guiding the Child,* by Rudolf Dreikurs and Margaret Goldman, classic Adlerian psychology, named after its founder Alfred Adler, suggests that reward and punishment as a means of child-rearing are faulty (Dreikurs & Goldman). In the book *The Courage to be Disliked,* the authors Fumitake Koga and Ichiro Kishimi do a wonderful job of illustrating this concept with the conversation between the two main characters, a philosopher and a young man. Koga and Kishimi use these characters to show how a reward-based model of discipline can be detrimental to an individual's development. With this model implemented, the individual will start to view a reward as his right, owed to him for the task of doing something right.

This effect, according to Koga and Kishimi, will have particularly negative effects when nobody is looking. The lack of praise in this model will result in the individual not participating in the right action if nobody is going to recognize them for it. If the motivation to do the right thing is distinctly tied to receiving positive reinforcements or a physical reward, how will the individual obtain these benefits with no observer present? Of course, the assumption is that the individual will eventually mature to the point that they understand doing the right thing simply because it's the right thing and *not* because they will receive a reward. However, this doesn't put the issue to rest quite yet because then we have to ask, “whose rules are they trying to play by”? What is particularly troubling is that reward and punishment methods of developing an individual’s conscience have been dominant throughout recent history. This raises a question: How does a tribe or society that uses reward-and-punishment-based methods of rearing coerce their people into doing the right thing even when nobody is looking? For a tribe that has its society’s rules entwined with their views on god and the divine, the solution would be to worship a god that is omnipresent. A god that says he is always watching, is always with you and knows even your deepest desires. This would be a highly effective way of coercing the individual into predictable and controlled behavior.

An omnipresent deity watching your every move is an effective method of influencing the person, but how does a group then create cohesion among many individuals? The group must establish a set of ground rules or societal norms that everybody must play by or else be exiled. This is the fundamental principle behind the ways our laws in the U.S. work (or are supposed to work). By creating a set of rules that the majority agrees will be the best route for society to take, then you can focus on the small group that defies these rules and implement a system to promote good behavior. Of course, that is an overly simplified explanation, but the point is simple — to have rules we can all agree to play by. This is where religion comes into the picture. For reasons that are beyond the scope of what we are discussing here, humans have a long tradition of creating gods and goddesses as a means to illustrate some aspect of reality that seems ineffable.

A religion that claims to be delivered from on-high by a superior being has the added benefit of distancing itself from the base level of human experience. It's much easier to defend the commandments of a deity that you cannot lay hands on or fight against. Consider the deity we just discussed, the one that is always watching to make sure you do the right thing. If that god seemed to only be concerned with whether or not *you* did the right thing but didn't seem to care much what other people did, then you would be in a fit of jealous rage. Therefore, a personal god that only concerned itself with what you specifically did would not be of much help to the group as a whole. But if it was explained to you that that deity makes you and all of your neighbors and friends play by the same rules, how much more trusting would you be of your fellow man? Knowing that society expects him to operate in a way your group is familiar with would make him seem more like family than simply another member of the tribe. There would then be a precedent and system in place to give order to daily behavior as well as provide consequences for violations of the norms.

This idea is related to concepts called the Moral Matrix and Moral Foundations Theory that is explored by Jonathan Haidt in his book *The Righteous Mind: Why Good People Are Divided by Politics and Religion*. Haidt explains how concepts like politics and religion relate to dispositions we hold from birth, and how these frameworks help create group cohesion. I won't attempt to do the brilliant concept justice by summing it all up in a few sentences. I would highly recommend reading the book in its entirety. However, to distill one crucial concept from Haidt's work, religion served and continues to serve as one of the strongest forces to create group cohesion that we have.

Let's now go back to the omnipresent deity. The idea that a god might be able to be at all places at all times, know all things, and still be operating within the micro-level context of the individual seems to be relatively new in the course of religion. Some of the oldest records we have describing humanity's early ideas of god illustrate what amounts to an absent father. The gods came out of the great, shapeless soup of creation and essentially subdivided the world until the ones responsible for the creation of things such as Earth arrived (Armstrong 9). However, these gods did not seem overly concerned with the daily events or decisions of their followers, much less were they omnipresent. Moreover, the understanding — according to the followers of the god Marduk — was that humanity was tasked with doing the work the gods did not have the time for (Mark). Humanity would, in turn, have its needs met so long as they heeded the demands of the gods. This meant that the gods would often be elsewhere, distant, occupied with some higher task that was out of the scope of human concern. We can see this tendency of the gods to be elsewhere — in other words, not omnipresent — in *The Akitu Chronicle* which describes how a specific festival honoring Marduk could not happen as planned because the statue of Marduk was gone (Mark). There are many accounts of rival tribes stealing the idols of their enemies and throwing that tribe into despair and disarray because their god was gone. Their god had literally left with the idol.

Imagine what kind of disturbance the introduction of an omnipresent deity who is also intimately concerned with the dealings of humans would create. It also would play on the moral foundations theory and create some of the strongest group cohesion humanity had seen. What has taken the stage in the theater of human religion is a god that is claiming to be above all other gods, is responsible for the creation of all things, has very specific rules for his followers, and knows your every move. With the creation of this deity, this tribe had solved the problem of personal accountability in private as well as the unification of group identity all within the framework of one single god. Creating a deity that has very specific rules and is so entwined with one's everyday life, as you might imagine, does come with consequences. Consider again the fact that reward and punishment methods of influencing an individual's behavior were dominant at this time.

What effect would the creation of this deity have on a society that operated on the reward-punishment framework? Perhaps the writer of this religion would be preoccupied with giving the god character speaking parts that seem bent on reward and punishment such as the threat of certain death as a consequence for not following a specific command from on high. Or the promise of a bountiful land as a reward for strict adherence to the rules, even if that bountiful land must be taken by force.

Once I began to understand the moral foundations theory, I was able to put words to one of my biggest concerns when it comes to the religion that I grew up in. This concern, I suspect, isn't limited to the Christian religion, but could be more broadly applied to other religious frameworks. Followers of gurus, cults, sages, deities, and the like are often guilty of this same mistake. My concern is that by adopting a religious framework, we aren't taking responsibility for our moral framework. One of the fundamental issues that religions face is what could be called "offloading" or perhaps, to borrow a term from psychology, deferment. The earliest instance of deferment or offloading is addressed in the concept of the social contract theory. In a distilled look, social contract theory is the idea that before humans began to form groups, they existed in a basic state of nature. However, to form cohesive, large-scale groups, humans had to enter into two fundamental agreements. The first agreement, called *Pactum*

Unionis, was essentially the responsibility for the individual to take on respect for another individual's property. This would allow for a sense of security among the members of the group or tribe.

The second agreement, more relevant to our discussion here, was *Pactum Subjectionis*. In this social contract, members agreed to imbue a person or group of people with the authority to enforce the regulations set out by *Pactum Unionis* (Laskar 1). Members of a tribe had to agree on a leader or leaders and agree among themselves to adhere to the rule of that leader. Some of the earliest "leaders" that a group selected was some form of deity, be it a physical person, a statue or idol, or an invisible entity. Regarding Julian Jaynes work that we referenced earlier, it is likely that the position of deferment went to whomever or whatever was attributed with the hallucinated voice of a god or previous leader.

For me to make my point clear, let's consider the Moral Foundations Theory and group evolution. As agriculture began to take root (pun intended), human beings started to become more group-ish. Groups competing with other groups for resources will naturally select for behaviors that help the group succeed. This means that traits or behaviors that promoted cohesion would be favored and those individuals would be most likely to reproduce. (This notion would fall under Darwin's theory of natural selection.) Through this evolution of group cohesion, an agreed-upon moral matrix would form and be used as the standard by which members of the group could determine whether loyalty to the tribe was being adhered to. We as humans created the moral structure, either explicitly or implicitly, that governed our behavior.

Whenever a group, tribe, or clan encounters an outsider or an "other" that seems to threaten any of the fundamentals of the moral foundations theory, humans will often start to use different language. We use distancing and powerful language to distinguish ourselves from the outsider. Drawing lines between us and them, and then placing ourselves above them on some aspect of morality. According to Dr. Rudolf Dreikurs in a 1958 publication called "The Cultural Implications of Reward and Punishment," placing ourselves above another is exactly what we do when we implement a reward and punishment model (Dreikurs). This illusion that we are something different from the "other" is the same one that supposes that we are the superior being to dish out rewards and punishments. When faced with someone who challenges our sense of identity, we might even create slang terms for the other so that mentally we can reconcile our hate against something we no longer see as equally human. We know this is the case. We've seen it throughout history with the struggles of every civil rights group demanding to be treated as equal humans. We even see it in the Bible when God, or perhaps the writer giving a voice to God, refers to another city with a different god as a "whore:" "How the faithful city has become a whore, she who was full of justice! Righteousness lodged in her, but now murderers" (ESV, Isa. 1:21).

In some verses in the Bible, there are even accounts of prophets telling some group, namely God's chosen people, to go into these cities where a different god is worshiped and kill every living thing. Raze homes and buildings and altars to the ground. The God character has decided that the society or race or city that doesn't adhere to His framework is less than, and He commands they be destroyed:

> "But the days are coming," declares the LORD, "when I will sound the battle cry against Rabbah of the Ammonites; it will become a mound of ruins, and its surrounding villages will be set on fire. Then Israel will drive out those who drove her out," says the LORD. (New International Version, Jer. 49:2)

The followers who hear and heed the message of the prophet then take up arms and, assuming the story is true, murder thousands of men, women, and children. Here we can see an example of offloading.

> Meanwhile, the remainder of the Jews who were in the king's provinces also assembled to protect themselves and get relief from their enemies. They killed seventy-five thousand of them but did not lay their hands on the plunder. (NIV, Est. 9:16)

Offloading occurs whenever an individual or group offloads the responsibility of defining morality into the hands of a seemingly supernatural entity. In these stories where the God character has demanded an entire population or race to be exterminated, those who heeded the call of the prophet have offloaded their morality into the hands of this prophet and the God he speaks for. This is where a potentially beneficial aspect of group and culture co-evolution has become malignant. It's evolved into a cancer. The responsibility has now become that of a supernatural being that is being interpreted by humans. Here is why this is a problem.

Think about the civil rights movements that we alluded to earlier. Throughout history, pick any of these cultural changes, and you will find groups using Biblical or religious texts to justify squashing this upheaval. You will also find groups using different parts of the same religious text to support the change. However, since the rights and liberties that many cultural revolutions fight for are not liberties mirrored in the Bible, for example, feminism or anti-slavery sentiments, it might be said that those using the text to squash the revolution may be truer to the text than we may like to admit. But here is the beautiful thing about this problem and the arc of history: as people take back the responsibility of defining morality, and then the larger culture must talk among itself about the newly agreed terms, we can make strides forward. You see, if we adhere to Biblical text, we are stuck in a morality matrix that was designed for a very different moment in human history. For example, should we continue to adhere to the Biblical law about promiscuous daughters?

> And the daughter of any priest, if she profanes herself by playing the harlot, profanes her father, she shall be burned with fire. (Revised Standard Version Bible, Lev. 21:9)

We would also be guilty of offloading the responsibility. We would be looking to an outside party to tell us what is right and wrong, and if anybody has a problem with it, we then defer to the seemingly supernatural entity as the authority. We adopt group identity and then commit unthinkable atrocities while feeling justified because they are in line with the rules of whom the group deferred to. We also give up our free will. The concept of offloading and free will are mutually exclusive. If we offload our responsibility to a deity then we also give up our agency to make decisions. But if we recognize that the group creates these matrices, we can perhaps move into a group conversation and create a cohesion that binds us together is our respect for our neighbors and freedom for ourselves.

So, What Now?

"It troubled me as once I was —

For I was once a child —

Deciding how an atom — fell —

And yet the heavens — held."

-Emily Dickinson

Simplicity played such a major role in the events that led to this book. Emily Dickinson's 4 lines of poetry whispered about the truth underneath my questions. I considered the honeybees and they chased away the god of my mind. The simplicity of this shift in thought is what made it so profound for me. I suspect that this is no coincidence. I think that the world around us holds the key to many questions the human mind has wrestled with since we first started to become social creatures. However, I don't think the world holds the keys in a guarded or cloaked manner. I think that they are in an open hand and it depends on what lens we are looking through that determines how we see the keys. I suspect that if we do our best to see simply what is before us, see what the world is saying, then we might find insights that make much more sense for our position in the universe. It seems that the outside world and the world we construct in our minds dance in a unique way. While this mental posture sounds bright and appealing, it is not without its consequence if you aren't careful. If you have had a similar life experience to mine, holding a particular world view or perhaps religion with a white-knuckle grip, then you may ask yourself where this leaves us. Where do we go once we let go of our crafted deities? When I first began to examine every idea, argument, and situation that I've discussed so far, I would sometimes find myself feeling alone. I was in the very position I just mentioned: feeling rather exposed in our world. That mental position is where we are going to start from in this chapter and show how changing the lens through which we see the same light might expose to us a deeper reality.

I've laid out one particular way to see things, but I haven't argued to close the Bible and disregard it entirely. What I'm suggesting is that perhaps we've misunderstood the nature of the god for which we seek, or, at the very least, have tried in vain to put God in a box that we can understand. As Armstrong illustrates in *A History of God*, humanity has gone through many iterations of expressions of the divine. While the Bible contains simply one iteration born of a specific time with specific pressures, it still gives us insight. It's a window into seeing the way that time shaped our current perceptions of divinity. That is why we don't need to shut the Bible and never open it again. We just need to see it through the appropriate lens, one of groups and cultures bumping up against one another and trying to make sense of the world around them. This is a perfectly reasonable tendency as it is one of the major ways that humans interact with and understand the world. We can see examples of this tendency in far more than just the way we think about God.

Consider discussions on political ideology or environmentally-motivated diets. When these topics arise between two individuals holding opposing viewpoints, nuance and subtle differences are glossed over and broad labels are applied to the other. We forget that every opinion that one might hold exists on a spectrum of views rather than inside one box of views or the other. So, to be able to understand our opponent or decide how to interact with them, we prematurely assign them a label — Democrat or Republican; Christian or Atheist; black, white, Asian, Hispanic — to comfortably interact. We tend to place others into boxes that we can easily understand due to preconceived notions that are often wrong or misinformed. When it comes to understanding God, we are no different in our ways.

This is a natural way to classify the world to feel safe by way of understanding one's surroundings. When you scale up the discussion from something like politics to the more nuanced discussion of God, it's clear that the pitfalls of this habit will scale up as well. When talking about God, we are trying to put into words the very thing that is beyond all words of all languages. This, I would argue, is why God is described in the Bible to be "like" certain things; like father or mother or a thief in the night. I'm simply pointing out that there could be much more to the nature of God than that. So don't throw out any hope or religious texts you might be in possession of.

When you start to realize that based on the interpretation of the Bible that we described earlier, God would be responsible for all the Good as well as all the Bad that we experience in this world, then you have to ask yourself, what then?

If God is both the holy and the profane, the left hand and the right, the light, and the shadow, what does that make Him?

In a lecture given by Quentin Skinner, he provides an elegant walkthrough of how we as a western society have progressed in our notion of freedom ("A Genealogy of Liberty: A Lecture of Quentin Skinner"). Progression not necessarily conveying moral progress so much as a movement along a path. Skinner does not claim to bring about one unifying concept of freedom, but rather leaves the definition of it up to the reader. His addition to the progression of ideas is that our idea of freedom could be predominantly based on not only the lack of interference in the pursuit of our interests but also that we are not dependent upon any other will or permission ("A Genealogy of Liberty"). This is a crucial distinction in understanding the concept of freedom. It doesn't just mean that there isn't anybody telling us what to do, forcing their views or policies onto our lives. It also means that we don't need anybody to supply us with such things. We can operate independently without requiring a specified framework. This view of freedom takes back the full weight of agency and places it squarely in the individual's hands.

In Adam in the room, we could argue that Adam has free will, but in Skinner's understanding, we could not argue that Adam is free even once the door is open. His environment is still controlled by someone else and he is dependent on the will of that power. How does this relate to our argument that God allowed evil into Eden to give Adam and Eve free will and thus an honest relationship with Him?

First, we must remember that nowhere in the scripture does it state that free will or honest love was God's motivation to allow temptation into the garden of Eden. Secondly, we must remember the choice that Adam and Eve are left with even after they are given free will: follow the righteous path of God or death. Skinner asserts that coercion by a negative, serious, and immediate threat means that one cannot truly choose ("A Genealogy of Liberty"). By this reasoning, after Adam and Eve are gifted free will, they aren't truly free because they are told that they will die if they eat of the Tree of Knowledge.

Aside from this idea within the framework of the creation story, what does it mean for how we relate to God in our daily life? If we accept Skinner's assertions, then are we truly free? What about the threat of Hell? Again, aside from the fact that our common notion of Hell isn't the idea argued in the Bible, does this notion allow us to be free? In the framework of coercion, Hell certainly is a negative and serious threat and one that is unavoidable. One of the foundational ideas in the story of Eden is that God gave Adam and Eve free will. Yet, since humans are inherently and naturally wired for self-preservation, why would they ever choose death? This serves to support Skinner's idea that Adam's and Eve's choice was not only easy but coercive. Further, let's examine the nature of the love that is being offered to Adam and Eve. True love means accepting a person for who they are. If love is anything less than that, it becomes a tool used to try to bring about change in the other. This logic means that true love has no opposite. It can't have an opposite. In the context of Skinner's ideas as they relate to this origin story, the threat of death, the promise of separation, the punishment, no way to return to that love, are all distinct opposites from the "true love" that has been offered.

If we are truly free so far as not being dependent on another's will or permissions, this means that nobody holds any power over us by way of dependency or coercion. What does this make God? Wouldn't this make God our equal?

In my experience, this question leaves you with two options. One: you can conclude that if God is everything in this sense, including both sides of the coin, then technically God is nothing. Kind of like if everybody is unique than nobody is. Do you see? If God is all that is, then there can be no "not-is" and therefore there is no real distinction between the two, making God not something separate from us. This, I feel, is a perfectly reasonable response to the question of what God is if He is both the good and the bad. What will it mean for your outlook on life? I can't be sure. However, whatever conclusion you come to is fair.

Secondly, you could conclude that this divine energy is not only running through all of creation down to the most subatomic particles but that it *is* the subatomic particle. You might look into the night sky and see as far as you can and contemplate the vastness of the universe and then turn around to look at the tiny dandelion that grew in the crack of the sidewalk and contemplate the electrons humming around in its bloom, only to realize that you are standing between these two monoliths of existence and are at the same time as close to them as a brother. When I think of all of this, I can see and feel that same consciousness that is within me. I can see it in that tiny, lonely dandelion, and in the brightest stars in the sky.

One aspect of this concept that seems to remain true, regardless of which of the above options you choose, is that the idea that we have free will can be both wrong and right at the same time. It just depends on your lens. It depends on the context of the conversation. For the sake of our interactions with each other on the individual level, it seems we do have a level of free will. I can choose, to some extent, how I am going to interact with someone I meet. This is a belief that is helpful to this context of interaction because it puts the responsibility of our behavior into our hands. It's a beneficial framework to operate within while at this level. However, we are still bound to the larger scale realities that we did not choose and cannot change. We were still born a product of two people we didn't pick. The presidential candidate we voted for lost. The family dog died of cancer. We are still operating within a larger framework that is not of our making and over which we don't have agency. Again we see how the world is infinitely more complex, more intertwined, more contradictory, and more beautiful than we see at first glance.

If we allow the book of Genesis to be what it is — a subversive counter-story of the nature of the world — then we must ask how the universe came into existence. If we take the best guesses science has to offer, we learn that the universe, not just the Earth, came from one singular point expanding and rushing out in all directions. This theory, in other words, is the Big Bang Theory. The term "big bang" is misleading because it causes us to picture an actual explosion, however, according to Deepak

Chopra, this is not the case (Chopra, Kafatos, 38). It misses the point because the idea of an explosion indicates that there must have been some material already in existence to explode and space within which the explosion could take place (Chopra, Kafatos, 38).

A more accurate image is there being nothing at all in the beginning. This may be a complicated "image" to hold in one's mind. The universe rushing into existence from nothing. This may seem even more unlikely than a literal explosion theory. To illustrate this idea, let's use an example.

Close your eyes and imagine you're in a room sitting at a table. Imagine an apple on the table. Now, imagine the apple turns into a donut and disappears. Then the table and the room disappear. Now it's just you surrounded by nothing. Now bring any kind of car into existence. Notice how the image of the car — color and make and model — essentially came from nothing? The paradigm of a car was served up to you by subconscious workings and was created in your conscious awareness from nothing that had already been there. The concept of the car arising in completeness without requiring construction is the main idea behind the universe arising from nothing. The idea mirrors ancient concepts that all the world is a dream being dreamt by some being. That dream would have arrived in the mind of that being from nothing. All at once.

Can you remember when you were a newborn and the thoughts you had then? No, you cannot. Can you remember the feeling of conscious awareness coming into your mind? Surely not. You can't remember anything before the awareness or the feeling of the awareness itself. The awareness and thus the ability to witness the world and create the mental reconstruction of the world came from nothing. It just appeared. And, let's not forget, that what we perceive of reality is only what our minds can reconstruct of reality.

I would argue that the universe coming into existence is much like a baby coming into awareness of itself and its surroundings. All at once, out of nothing, the construction of reality is born. Something being created from nothing happens all the time, all around us. Subatomic particles blink in and out of existence, yet we don't know where they come from, or where they go. When specific factors are present, an atom may gain or lose an electron. This causes an electron to jump from one shell of orbit to another. The trick is, when the electron makes the change from one shell to another, it doesn't travel the distance in between. In *You Are the Universe,* the authors explain that the electron blinks out of existence and then reappears in the new orbit (Chopra, Kafatos, 90). The fundamental building blocks of our reality are rushing into existence from nowhere. In this way, they are carrying on the same dance that started this grand play; appearing from nothing, all at once. The universe is more like the dream of God in this way. Not that we are all a dream of some other being, but that everything *is* this being. Down to the basic building blocks

It's been said that human beings bring context to the universe. In reality, there are no colors like blue and green, no sounds like harmonious singing, and no smells like warm freshly made bread. This is brought into existence through the conscious filter of the human mind. At this point, I would like to point out that I am not arguing that we humans are the creators. I think it's bigger than that.

Take, for example, the double-slit experiment. If you are unfamiliar with the test, to put it roughly, scientists were attempting to determine if light behaved like a particle or a wave. By firing electrons through a single slit, and then a double slit, they hoped to find patterns to indicate what the behavior of the electrons were. To their astonishment, the results proved both hypotheses. Light seemed to behave like a particle *and* a wave. It gets stranger. Whenever these scientists would observe the particles passing through the slits, the particles would seemingly choose to behave like a particle *or* a wave. The very act of human observation brought about a specific reality (Thompson).

Let's go deeper with a question. Why does your body stay together on the atomic level? Of course in subatomic relationships, atoms by way of their electrons maintain a consistent material such as your skin or hair. But that's just the how not the why. Why do these atoms and electrons stay together right now, but then break apart later? What kind of — to borrow a term from comedian Pete Holmes — conspiracy of electrons do we have here? What do the implications of the double-slit experiment and the role of the observer mean for the world around us? If our conscious observation of the atoms locks them into a specific state, just like a wave or a particle, does this imply that some kind of consciousness is locking our reality into a specific arrangement? Do we exist because we know we exist? It may be difficult to contemplate, but taking into consideration the results of the double-slit experiment, why then shouldn't we apply that same logic and those same results to our existence? After all, at our basest level, we are made of electrons.

At the very basic building block level of reality, there seems to be a personality within every piece of it. You see, I don't think we should stop at asserting that human observation brings about a specific reality. There is an awareness, a personality, at the foundation of everything and we happen to be one of the ways this creation gets to look at itself, hear itself, taste itself, and fully enjoy the marvel of being. One could say am-ness or simply, I Am.

As lower-level beings create the higher-order beings, such as the worker bees creating the hive-mind or our neurons creating or receiving consciousness, this cycle continues. Single-cell organisms making up multi-cellular ones that make up an organ or tissue. These tissues and organs make up a creature, be it mammal, reptile, amphibian, insect, plant, or homo sapiens. The question we are on the cusp of answering is, "Where does the cycle go from here?" What larger entity does the collection of humans make? Perhaps we are organisms making up a larger system on our planet much like how a cluster of atoms makes an element. The point here is that this cycle can keep going — humans to planets to galaxies and so on. If we can agree that these cycles continue to ripple out as a result of numerous organisms making up larger systems, it would follow that the cycle will continue to ripple out so long as there are groups of organisms left to make up the next level. The only point at which this cycle would cease to expand would be when it reaches one single entity that is made up of all previous levels.

Even after we consider that we may be intimately linked with our world, each other, and the broader universe, a question seems to remain. Even after we decide to face the truth about good and evil we must consider why are we asking these questions? What inside us makes us uneasy or unsettled to the point that we will construct giant narratives about how we came to be and who is pulling the strings? Yes, good and evil must exist as they are two sides of the same coin. Yes, our religious narratives do hold logical inconsistencies and are a product of their time and of the people of that time. But even if we decide to adopt this approach to seeing the world, by accepting the role of good and bad to be sober in our relation to our environment, why do we have to work and strive to operate in this specific way? I suspect we are aware of no other creature that struggles in the way humans do with the mental framework or schema with which they operate. Other creatures don't seem to stir and question and squirm in the way we humans do. What is that special ingredient in us?

I don't think I can come to a definitive answer to this question. At least, not one that seems to satisfy the wondering in my mind, much less anyone else's. However, consider where we have gone in this book. When we begin to look at the very real way that our consciousness seems to be connected to our environment, we are almost taken back to a spiritual position on human beings. Back to where we started, yet, at a bigger perspective. There is a direct relationship between consciousness and that which is observed by the consciousness. Studies and studies have shown this. On an individual level, concerning how one conscious entity relates to their environment, this research can give us a way of understanding this relationship.

What has this book been? It has been my attempt at explaining the way I see things most honestly and genuinely as I could. We discover things as we grow older. By that I don't mean simply learning new things, I mean discover them like looking behind the veil, learning the sleight of hand that lets the magic work, or uncovering secrets. My life has been a continuous path of discovering the ideas and people I placed on a pedestal, the idols I created and venerated in my mind, and the God that I chased after. I have had an unquenchable drive to poke and dig and push against my foundations for as long as I can remember. Perhaps growing up in a Christian school, one that I opposed in almost every way, set me in a disposition to question the occupants of all ivory towers. This may sound disheartening or sad at the start, but what it did was force my understanding and capacity for love to grow. By taking apart the walls I had built using religion and "divinely-inspired" world views, I was able to see how much bigger my world was. I could see how my life was part of the same vein, branch, vine as every other person I met. I considered the way I felt about God — the holiness He had, the fear it held in me, and the immense trust I placed in his hands. When I started to push back against the texts that told this God's story, I saw that He seemed to be the mouthpiece for angry men who were doing their best to survive and thrive, but betrayed their humanity and anger and weakness all the same. I discovered God the way climbers discover mountains. It's only once you recognize that the mountain can only hold the amount of power over you that you are willing to give up, that you can climb that mountain. What happens when you take your power back? What are you left with when your fears, your inhibitions, your mountain, and your God can no longer use intimidation as leverage against you? You are left with the feeling of your true self and that is a fantastic feeling. You take back that power. Go get it.

That's all this book has been. My understanding and assumptions of how reality works have grown and changed so much and will continue to do so. And in this, I have felt hope. Not a hope specific for some future event, but a kind of joyous, in-the-moment hope. Hope for the day. I hope that while you read this, you can feel your shoulders relax and fall, your tongue ease from the roof of your mouth, your abdomen soften and breathe deeper knowing that at the very least, life is a lot bigger than our categories suggest. I want you to know that it's beautiful and terrifying, ugly and divine, and you are not alone in any of it. You, an incredibly complex organism made up of human and non-human cells, a creature that can express their inner voice in any number of ways, a magnificent arrangement of electrons that can inspire, hold, and care, a perfectly produced fruit of the earth that can steward or destroy, are simply breathtaking. There is no limit to the wonder and insight we hold. Please let your grip on how the world works soften just a little.

Works Consulted

@RichardDawkins (Richard Dawkins) "Do dogs go to Heaven? Only pets, not wild animals? If only humans go to Heaven, how far back? Homo erectus? Australopithecus? Only people who lived after Jesus? Not Abraham, Moses or Elijah, then. Which 'you' will be in Heaven? Senile you with dementia? Midlife-crisis you?" *Twitter*, 26 Mar. 2018, 4:47 a.m., twitter.com/richarddawkins/status/978191723850498049?lang=en.

"21.24." *Mysterious Universe,* 22 June 2019, https://mysteriousuniverse.org/2019/06/21-24-mu-podcast/

"A Genealogy of Liberty: A Lecture by Quentin Skinner." *YouTube,* uploaded by Stanford, 1 Dec. 2016, https://www.youtube.com/watch?v=PjQ-W2-fKUs&t=1679s.

"Abelard's Ethics for Today." *YouTube,* uploaded by Holy Spirit Moments, with Fr. Kerry Walters, 9 May 2018, https://www.youtube.com/watch?v=UgWDYWKSeVc.

"Africanized Bees." *Smithsonian Institution*, https://www.si.edu/spotlight/buginfo/killbee

"Bee Still My Heart." *Outside Podcast*, 5 December 2017, https://www.outsideonline.com/2266981/bee-still-my-heart

"Brain Area Unique to Humans Linked to Cognitive Powers." *University of Oxford*, 28 Jan. 2018, http://www.ox.ac.uk/news/2014-01-28-brain-area-unique-humans-linked-cognitive-powers.

"From Tree to Shining Tree." *Radiolab*, 30 July 2016, https://www.wnycstudios.org/podcasts/radiolab/articles/from-tree-to-shining-tree

"Pollination Facts: Honey Bees Are Pollinators." *American Beekeeping Federation*, https://www.abfnet.org/page/PollinatorFacts.

"The Global Consciousness Project – Ph.D. Roger Nelson." *YouTube*, Uploaded by Exopolitik, 3 Feb. 2013, https://www.youtube.com/watch?v=HQXoMI130KY.

"What is INSTINCT? What does INSTINCT mean? INSTINCT meaning, definition & explanation." *Youtube*, uploaded by The Audiopedia, 6 November 2016, https://www.youtube.com/watch?v=BIk3RsjxpNo&t=6s

Abelard, Peter, and Paul Vincent. Spade. *Ethical Writings: His Ethics or "Know Yourself" and His "Dialogue between a Philosopher, a Jew and a Christian*. Hackett Publ. Co., 2006.

Abramowitz, Jack. "*22. Nefesh and Neshama - HaShoneh Halachos 2: Mishneh Torah." OU Torah*, 22 July 2015, https://outorah.org/p/25685.

AnticitizenX. "The Problem of Omnipotence." *YouTube*, 21 February 2018, www.youtube.com/watch?v=gNQkSJXUzjo&list=PLK9PFKVP1h7Y3HdZhzhZk6CeaF-qDJGlQ&index=4&t=0s

Armstrong, Karen. *A History of God*, Vintage, 1999.

Barker, Dan, and Richard Dawkins. *God: The Most Unpleasant Character in All Fiction*. Sterling, 2016.

Bronstein, Michael V., et al. "Belief in Fake News Is Associated with Delusionality,

Dogmatism, Religious Fundamentalism, and Reduced Analytic Thinking." *Journal of Applied Research in Memory and Cognition*, vol. 8, no. 1, 17 May 2018, pp. 108–117, doi:10.1016/j.jarmac.2018.09.005.

Cammaerts, Marie-Claire, and Roger Cammaerts. "ARE ANTS (HYMENOPTERA, FORMICIDAE) CAPABLE OF SELF RECOGNITION?" *Journal of Science*, vol. 5, no. 7, 2015, p. 521., doi:10.21276/jos.

Chittka, Lars, and Jeremy Niven. "Are Bigger Brains Better?" *Current Biology*, vol. 19, no. 21, 17 Nov. 2009, pp. R995-R1008, doi:10.1016/j.cub.2009.08.023

Chopra, Deepak, and Menas Kafatos. *You Are the Universe: Discovering Your Cosmic Self and Why It Matters.* Random House UK, 2018.

Dickinson, Emily. "It troubled me as once I was—." *American Poems - Analysis, Themes, Meaning and Literary Devices*, https://www.americanpoems.com/poets/emilydickinson/it-troubled-me-as-once-i-was/.

Dolansky, Shawna. "How the Serpent Became Satan." *Biblical Archaeology Society*, 8 June 2019, www.biblicalarchaeology.org/daily/biblic

al-topics/bible-interpretation/how-the-serpent-became-satan/.

Dreikurs, Rudolf. "The Cultural Implications of Reward and Punishment." *International Journal of Social Psychiatry*, vol. 4, no. 3, pp. 171–178, 1 December 1958, doi:10.1177/002076405800400302.

Dreikurs, Rudolf and Margaret Goldman. *The ABC's of Guiding the Child.* North Side Unit of Family Education Association, 1967.

Drummond, John G. "Who Is Satan?: The Many Forms of the Devil in the Bible." *Biblical Archaeology Society*, 10 Sept. 2019, www.biblicalarchaeology.org/daily/biblical-topics/bible-interpretation/who-is-satan/.

Haidt, Jonathan. *The Righteous Mind: Why Good People Are Divided by Politics and Religion.* Penguin, 2013.

Hammer, Jill. "Lilith: Lady Flying in Darkness." *My Jewish Learning*, https://www.myjewishlearning.com/article/lilith-lady-flying-in-darkness/. Accessed 3 January 2020.

Harris, Sam. *Free Will*. Free Press, 2012.

Hitchens, Christopher. *Mortality*, New York : Twelve, 2012, pp. 21–22, www.archive.org.

Howell, Elizabeth. "How Long Have Humans Been on Earth?" *Universe Today*, 23 Dec. 2015, https://www.universetoday.com/38125/how-long-have-humans-been-on-earth/. Accessed 3 January 2020.

Jaynes, Julian. *The Origin of Consciousness in the Breakdown of the Bicameral Mind*. Kindle. Houghton Mifflin, 2000.

Jozuka, Emiko. "Aboriginal Australians Are Earth's Oldest Civilization: DNA Study." *CNN,* 22 Sept. 2016, https://www.cnn.com/2016/09/22/asia/indigenous-australians-earths-oldest-civilization/index.html. Accessed 25 November 2019.

Kishimi, Ichiro, and Fumitake Koga. *Courage to Be Disliked: How to Free Yourself, Change Your Life and Achieve Real Happiness,* Allen & Unwin, 2018.

Laskar, Manzoor. "Summary of Social Contract Theory by Hobbes, Locke, and Rousseau." *SSRN Electronic Journal,* 4 Apr. 2013, doi:10.2139/ssrn.2410525.

Leucht, Stefan, et al. "How Effective Are Common Medications: A Perspective Based on Meta-Analyses of Major Drugs." BMC Medicine, vol. 13, no. 1, 2015, p. 3. Crossref, doi:10.1186/s12916-015-0494-1.

Levenson, Jon D. "Genesis: Introduction and Annotations". The Jewish Study Bible, Edited by Adele Berlin and Marc Zvi Brettler. Oxford University Press, 2004.

Mark, Joshua J. "The Marduk Prophecy." *Ancient History Encyclopedia*, 14 Dec. 2016, https://www.ancient.eu/article/990/the-marduk-prophecy/. Accessed 4 Jan. 2020.

Meyers, Deborah et al. "Relationship of Obesity and Physical Fitness to Cardiopulmonary and Metabolic Function in Healthy Older Men," *Journal of Gerontology*, vol. 46, no. 2, pp. M57-M65, March 1991, https://doi.org/10.1093/geronj/46.2.M57

Michener, Charles D., and David A. Grimaldi. "The Oldest Fossil Bee: Apoid History, Evolutionary Stasis, and Antiquity of Social Behavior." *Proceedings of the National Academy of Sciences*, vol. 85, no. 17, Jan. 1988, pp. 6424–6426., doi:10.1073/pnas.85.17.6424. Accessed 24 Nov. 2019.

Montgomery, Sy. *The Soul of an Octopus: A Surprising Exploration into the Wonder of Consciousness.* Simon & Schuster UK Ltd., 2016.

Morris, John D. "Does the Bible Really Claim That Insects Only Have Four Legs?" *Institution for Creation Research,* 1 July 200.1 https://www.icr.org/article/does-bible-really-claim-that-insects-only-have-fou

Narby, Jeremy. *Intelligence in Nature: An Inquiry into Knowledge.* Jeremy P. Tarcher/Penguin, 2006.

Plath, Sylvia. "Mad Girl's Love Song," 1953.

Popova, Maria. "T.S. Eliot on Writing: His Warm and Wry Letter of Advice to a Sixteen-Year-Old Girl Aspiring to Become a Writer." *Brain Pickings,* https://www.brainpickings.org/2018/01/18/t-s-eliot-alice-quinn-letter/. Accessed 18 Jan. 2018.

Public Library of Science. "Fossil Record Reveals Elusive Jellyfish More Than 500 Million Years Old." *ScienceDaily,* 2 Nov. 2007, www.sciencedaily.com/releases/2007/10/071030211210.htm. Accessed 1 March 2019.

Quammen, David. *Tangled Tree: A Radical New History of Life*. William Collins, 2019.

Radin, Dean I. *Entangled Minds: Extrasensory Experiences in a Quantum Reality*. Pocket, 2006.

Radin, Dean I. *Real Magic*. Harmony, 2018.

Radin, Dean, and Roger Nelson. "Evidence for Consciousness-Related Anomalies in Random Physical Systems." *Foundations of Physics*, vol. 19, no. 12, 1989, p. 8, www.cia.gov/library/readingroom/docs/CIA-RDP96-00789R002200520001-0.pdf.

Radin, Dean I., and Marilyn J. Schlitz. "Gut Feelings, Intuition, and Emotions: An Exploratory Study." *The Journal of Alternative and Complementary Medicine*, vol. 11, no. 1, 2005, pp. 85–91., doi:10.1089/acm.2005.11.85.

Reeves, Ryan. "William of Ockham." *YouTube*, 2 Aug. 2015, https://www.youtube.com/watch?v=B6y-3tiwO5Y&t=1505s.

Seeley, Thomas D. *The Wisdom of the Hive: The Social Physiology of Honey Bee Colonies*. Acme Bookbinding, 2008.

Stewart, Don. "When Did Moses Write, or Compile, the Book of Genesis?" *Blue Letter Bible,* https://www.blueletterbible.org/faq/don_stewart/don_stewart_678.cfm.

Suzuki, David. "1 Of Every 3 Bites of Food Depends on Bees – Let's Save Them." *HuffPost Canada, HuffPost Canada,* 1 Sept. 2014, www.huffingtonpost.ca/david-suzuki/colony-collapse-disorder-_b_5549504.html.

The Bhagavad Gita. Translated by Franklin Edgerton. Harvard University Press, 1997.

The Bible. New International Version, Zondervan, 2005.

The Bible: New Revised Standard Version, Thomas Nelson Publishers, 1989.

The Bible. King James Version, Holman Bible Publishers, 1973.

Thompson, Avery. "The Logic-Defying Double-Slit Experiment Is Even Weirder Than You Thought." Popular Mechanics, Popular Mechanics, 15 Feb. 2018, https://www.popularmechanics.com/science/a22280/double-slit-experiment-even-weirder/.

Tylenol. “Tylenol® and Your Aspirin Use.” *TYLENOL®*, www.tylenol.com/safety-dosing/health-conditions/heart-health. Accessed 11 Apr. 2020.

Walters, Kerry. “Abelard’s Ethics For Today.” *YouTube*, 9 May 2018, https://www.youtube.com/watch?v=UgWDYWKSeVc&t=1s

Waltke, Bruce. “The Creation Account in Genesis 1:1-3.” *Dallas Theological Seminary, Bibliotheca Sacra*, vol. 132, 1975, pp. 25-36, https://faculty.gordon.edu/hu/bi/ted_hildebrandt/otesources/01-genesis/text/articles-books/waltke-cosmogony-bsac.pdf. Accessed 5 Jan. 2020.

“ What Is a Z Score What Is a p-Value.” *What Is a Z Score What Is a p-Value*, resources.esri.com/help/9.3/arcgisdesktop/com/gp_toolref/spatial_statistics_toolbox/what_is_a_z_score_what_is_a_p_value.htm.

Witruk, Evelin, Angela D. Friedrici & Thomas Lachmann, editors. *Basic Functions of Language, Reading and Reading Disability.*

Springer Science Business Media, LLC, 2002.

Wolf, Maryanne *Proust and the Squid: The Story and Science of the Reading Brain*. E-book, Harper Perennial, 2017.

About the Author

Justin Herb spent most of his childhood in the small town of Medora, Indiana. From there his family moved to Columbus, Indiana where he attended high school. Most of his childhood was spent in a private Christian school and attending the associated church. This is the place where early questions began to draw his attention and set him on a path of investigating and deconstructing his faith. The drive to pull apart ideas and see what they are made of has propelled Justin through many obstacles life has thrown his way, and has led to the creation of this book. He aims to untie the knot that many hold due to questions about the faith they hold. He hopes that this book will help the reader to feel at ease, to know the questions are good and part of the process and that our reality is so much bigger than our categories can hold.

https://godofhoneybees.com

https://godofhoneybees.weebly.com/email-signup-form.html

Made in the USA
Monee, IL
05 November 2026

67383323R00097